Nanofabrication Techniques - Principles, Processes and Applications

Edited by Dipti Ranjan Sahu

Published in London, United Kingdom

Nanofabrication Techniques - Principles, Processes and Applications
http://dx.doi.org/10.5772/intechopen.107730
Edited by Dipti Ranjan Sahu

Contributors
Thangavel Krishnasamy, Balaprakash Vadivel, Prema Rangasamy, Mahitha Mohan, Ahmed Hashem Abdelmohsen, Sherif A. El-Khodary, Nahla Ismail, Divya Gupta, Rimpi Kumari, Amena Salim, Rahul Singhal, Sanjeev Aggarwal, Arnab Ganguly, Gobind Das, Dipti Ranjan Sahu, Yu-Min Shen, Jow-Lay Huang, Sheng Chang Wang, Kai-Ming Hu, Wen-Ming Zhang, Payam Hayati, Saeedeh Ebad-Sichani, Mohammad Jaafar Soltanian-Fard, Paria Sharafi-Badr, Vishal Pande, Sachin Namdeo Kothawade, Sharmila Kuskar, Dinesh Chakole, Sandesh Bole

First published in London, United Kingdom, 2023 by IntechOpen
IntechOpen is the global imprint of INTECHOPEN LIMITED, registered in England and Wales, registration number: 11086078, 5 Princes Gate Court, London, SW7 2QJ, United Kingdom

British Library Cataloguing-in-Publication Data
A catalogue record for this book is available from the British Library

Additional hard and PDF copies can be obtained from orders@intechopen.com

Nanofabrication Techniques - Principles, Processes and Applications
Edited by Dipti Ranjan Sahu
p. cm.

This title is part of the Nanotechnology and Nanomaterials Book Series, Volume 2
Topic: Nanotechnology and Nanodevices
Series Editor: Jung Huang
Topic Editors: Sadia Ameen and M. Shaheer Akhtar

Print ISBN 978-1-83769-048-0
Online ISBN 978-1-83769-049-7
eBook (PDF) ISBN 978-1-83769-050-3
ISSN 3029-0538

For EU product safety concerns:
IN TECH d.o.o., Prolaz Marije Krucifikse Kozulić 3, 51000 Rijeka, Croatia,
info@intechopen.com or visit our website at intechopen.com.

IntechOpen Book Series

Nanotechnology and Nanomaterials

Volume 2

Aims and Scope of the Series

Humans face growing challenges in environmental sustainability. We need better materials, more effective devices, and higher computing power to conquer these challenges. Researchers have long explored the realm of science at the nanometer scale for developing novel problem-solving technologies. In the past decades, breakthroughs in multiscale computer simulation, novel hierarchical material and device structures, nanometrology, and new functionalities from interfacial and quantum size effects have enhanced our ability to utilize nanotechnology to solve real-world problems. This book series addresses important advancements in nanotechnology and nanomaterials. It showcases how nanotechnology is being continually developed and implemented in a variety of domains of science and technology. The two main topics this series covers are Nanotechnology and Nanodevices, and Nanomaterials and Nanostructures.

Meet the Series Editor

Jung Y. Huang, a university educator and researcher, has been working to unravel the structures and functional properties of materials and cellular events in living cells with varying optical methodologies. He has co-authored hundreds of journal papers and five book chapters. He also holds tens of patents in laser techniques and single-molecule/hyperspectral imaging and has developed architectural photonics based on hierarchically structured materials. Currently, his research focuses on the integration of artificial intelligence methodology with optics to automatically discover meaningful information from optical sensing/imaging data cubes. He has been an editorial board member and reviewer for several scientific journals. As a member of the global scientific community, he sincerely supports and endeavors to promote the spread of scientific knowledge.

Meet the Volume Editor

Prof. Dipti Ranjan Sahu is a Professor of Physics in the Department of Biology, Chemistry and Physics at the Namibia University of Science and Technology (NUST). He received his Ph.D. in Physics from the Institute of Materials Science, Utkal University, India. After obtaining his Ph.D., he worked as a postdoctoral researcher and visiting scientist/faculty at several institutions, including National Taiwan University, National Cheng Kung University, Taiwan, and the University of Witwatersrand, South Africa. His research focus is on multifunctional materials including nanomaterials, ceramics, composites, spintronics, and magnetic materials, as well as the application of functional materials in devices. He has published more than 115 peer-reviewed articles, 4 books, 4 book chapters, and more than 100 research articles in conference proceedings and meetings.

Contents

Preface

Nanofabrication is the process of developing different nanostructures that are commonly used as components of systems or devices. Advances in nanofabrication technologies have produced various kinds of nanostructures, allowing for precise control over size, shape, heterogeneity, and material composition. Recently, nanofabrication methods have been receiving increasing interest due to their importance for the creation of compact devices and their assembly into functional systems for different applications in nanoelectronics, semiconductors, optics, nanophotonics, biomedical engineering, bio-inspired architectures, and nanofluidics for the production of high-value-added products.

New techniques and hybrid nanostructures are developed using different synthesis and fabrication procedures. Different processes, phenomena, and interesting properties are studied theoretically and experimentally using advanced characterization and analysis techniques. Nanofabrication techniques are the building blocks of different devices and technological innovation. Advances in modern nanofabrication techniques contribute to state-of-the-art implementations in various areas. Nanofabrication techniques will always benefit from further development regarding structural, mechanical, optical, and magnetic or electronic properties of the processing materials for device applications.

This book presents an overview of the working principles, recent progress, and the challenges and limitations in the fabrication of nanostructures. It also discusses the basic phenomenon of nanostructure formation and the different types of nanostructures and their synthesis, fabrication, application, and devices. The book provides useful information on nanofabrication for students, researchers, and professionals. The latest research on different types of nanostructures, scientific advances in fabrication techniques, and exciting potential applications of nanostructures will hopefully spur the development of future devices. The book addresses the scientific advances and interesting facts of nanofabrication for the study of next-generation devices. Finally, it examines the technological implications of further advances and presents a perspective on the future direction of large-scale nanofabrication.

This book contains eight chapters that explore some of the most exciting developments of nanostructure materials, fabrication techniques, and applications. Chapters 1 and 2 discuss nanofabrication methods and the basics of fabrication of different nanostructures. Chapters 3, 4, and 5 discuss various morphologies, mesoporous silica, and nano ZnO. Chapters 6, 7, and 8 review assembly technology of different materials, including 2D materials, and basic performance application of materials. These chapters explain the fundamentals of nanostructure materials and fabrication techniques for technological innovation.

I would like to thank the contributing authors for their excellent chapters. I would also like to thank the staff at IntechOpen for their help throughout the publication process.

Dipti Ranjan Sahu
Faculty of Health, Natural Resources and Applied Sciences,
School of Natural and Applied Sciences,
Department of Biology, Chemistry and Physics,
Namibia University of Science and Technology,
Windhoek, Namibia

Section 1

Nanofabrication Method

Chapter 1

Nano-Fabrication Methods

Saeedeh Ebad-Sichani, Paria Sharafi-Badr, Payam Hayati and Mohammad Jaafar Soltanian-Fard

Abstract

In the field of investigating nano-fabrication, it is not possible to reach a single and separate definition compared to macro-fabrication. Nano-fabrication can be defined as an assembly process to produce a one-dimensional, two-dimensional, or three-dimensional structure at the nanometer scale. The importance of recognizing and examining nanofabrication techniques considering the revolution that nanofabrication compounds have in molecular adsorption, catalysis, magnetism, luminescence, nonlinear optics, and molecular sensing, have been known because they provide the possibility of reproducible mass production in this field. In this chapter, to create a general understanding of nano-fabrication and the challenge of creating nanometer size reduction, we will review new tools and techniques for the production of nanostructures, which are divided into three major parts: thin film, lithography, and engraving.

Keywords: nano-fabrication, micro-fabrication, lithography, thin films, etching, pharmaceutical, medical fields

1. Introduction

Nanotechnology has already found numerous applications in various fields, including electronics, medicine, energy, and environmental remediation. For example, nanoscale materials fabricate electronic devices like transistors and solar cells. In medicine, nanoparticles are used for drug delivery and imaging. Nanotechnology is being explored in energy to develop more efficient batteries and solar cells. In environmental remediation, nanoparticles are used to remove contaminants from air and water.

In conclusion, nanotechnology has the potential to revolutionize many fields and improve our lives in numerous ways. However, it is essential to approach this technology cautiously and continue to study its potential risks and benefits. With responsible development and use, nanotechnology can be a powerful tool for addressing many of humanity's challenges.

Despite the many benefits of nanotechnology, there are concerns about the possible dangers of using nanomaterials. These risks include toxicity, environmental impact, and ethical considerations. As such, it is important to continue to study the advantages and disadvantages of nanotechnology and to develop appropriate regulations and guidelines for its use.

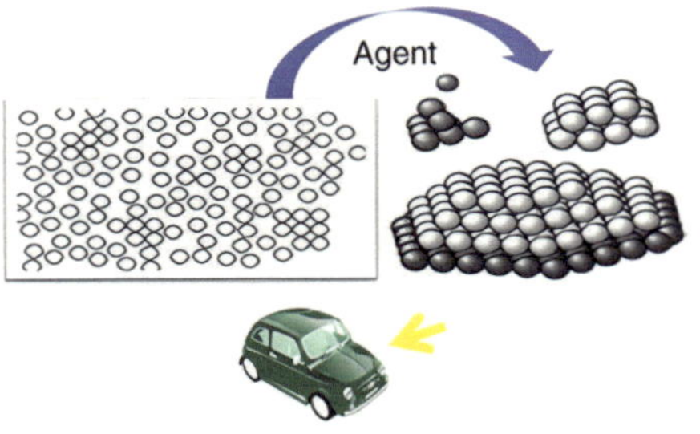

Figure 1.
Cartoon representation of Feynman's atom-by-atom [1].

One way to address these concerns is through responsible research and innovation (RRI) frameworks that integrate ethical, social, and environmental considerations into developing and deploying new technologies. Additionally, transparency and communication with stakeholders, including the public, are crucial in ensuring that the benefits of nanotechnology are realized while minimizing its potential risks. By taking a proactive and collaborative approach, we can ensure that nanotechnology is used safely and responsibly, benefiting society.

Richard P. Feynman, a Nobel Prize-winning physicist, spoke about the potential of molecular-scale engineering in " There is Plenty of Room at the Bottom" on December 29, 1959, at the California Institute of Technology. Nobel laureate Richard Feynman is frequently credited with the "birth of nanotechnology." In his creative speech, Feynman described a nano world in which atoms might be ordered individually. Feynman awarded monetary incentives to individuals who could complete his nano tasks. Despite being a visionary, Feynman overlooked the significance of chemistry in developing nanotechnology. He proposed that changing and controlling individual atoms and molecules would be possible, creating novel compounds, electronics, and medical treatments. This vision has become a reality with the advent of nanotechnology, which allows scientists to work at the atomic and molecular levels.

Figure 1 is a cartoon representation of Feynman's atom-by-atom (ABA) fabrication technique for creating objects like cars from a single atom [1, 2].

The Birth of Nanotechnology is Atom-by-Atom Manufacturing. The origin of "nano" is the Greek word *ναvoζ*. The metric prefix "nano" indicates 10^{-9}, and 1 nm (one nanometer) is 10^{-9} m [2].

The term "nano" derives from the Greek word o. The metric prefix "nano" is derived from the Greek o, metro, which means "unit of measurement."

2. Definition of nano-fabrication

Nano-fabrication" is a fabrication technology used in nanotechnology. Nano-fabrication is employed for fabricating components more minor than the ones fabricated by microtechnology, generally smaller than 1 μm, and the definition of nano-fabrication is unclear. One of the merits of this technology is that scientists have reached the theoretical limit of accuracy, for example, the size of a molecule or an atom; another belief is that we are dealing with a "super technology" [1].

Microtechnology has been transformed into nanotechnology with slow changes, and many laboratories changed the name of their microfabrication laboratories to nano-fabrication laboratories, while there were no fundamental changes. It should be remembered that politics and economics partially caused these changes. With the changes and progress in the micro-scale, we slowly reached the nanoscale, and then its name was also changed [3].

Nano-fabrication is critical in advancing several areas, including electronics, biotechnology, energy, and materials science. It has enabled the production of smaller and more efficient devices, such as nanosensors, nanoelectronics, and nanomedicines. The techniques used in nano-fabrication include lithography, self-assembly, molecular beam epitaxy, and sol-gel synthesis. Each methodology has benefits and drawbacks, and the choice of the method relies on the desired application. Nano-fabrication also poses challenges, such as the need for high precision and accuracy, the control of surface properties, and the scalability of the process. However, with advancements in technology and materials science, these challenges are being addressed, and nano-fabrication is becoming more accessible and widespread. Overall, nano-fabrication is a powerful tool that has revolutionized many fields and has the potential to continue driving innovation and progress in the future.

3. Special properties of nanomaterials

The two main reasons for changing nanomaterials' chemical and physical properties are surface effects and entering the world of quantum physics. The meaning of surface effects is the increase of the surface area-to-volume ratio by decreasing the particle size and reaching dimensions below one hundred nanometers. Increasing the proportion of surface atoms in the material causes the properties of the surface atoms to affect the properties of the whole material. Among these effects, very high reactivity can be mentioned. Another reason is entering the world of quantum physics and dissociating energy bands and turning them into energy levels. Due to this factor, special optical properties are observed in quantum dots or quantum wires that exhibit ballistic electrical conductivity. In this article, these two reasons are stated and analyzed [4].

The second reason for the unique properties of some nanomaterials is entering the world of quantum physics and dissociation of energy levels. First of all, it should be stated that this phenomenon is not valid for all nanomaterials, and it only happens for quantum nanomaterials. Quantum nanomaterials are compounds whose one, two, or three dimensions are less than the critical dimensions required to enter the world of quantum physics. The critical dimensions necessary for nanomaterials are determined according to their type. The critical dimensions for semiconductors are about 10 nm, and for conductors, about 1–3 nm [5, 6].

As the particle becomes smaller and its dimensions reach below one hundred nanometers, the increase in the percentage of surface atoms becomes significant. As the size decreases, the slope of the rise in the ratio of surface atoms also increases. **Figure 2** shows the particle size reduction's effect on the surface atoms ratio (surface area-to-volume ratio). As can be seen, as the particle becomes smaller and its dimensions reach below one hundred nanometers, the increase in the percentage of surface atoms becomes significant. With the decrease in size, the slope of the rise in the proportion of surface atoms also increases [4].

Materials can be divided into semiconductors, insulators, and metals or conductors. The qualities of a material may be changed at nanoscale length scales, which affects how its electrons behave (delocalized, localized, or somewhere in between). For instance, when a metal gets smaller, its electrons are confined to specific energy levels rather than being free to travel through energy bands. The characteristics of

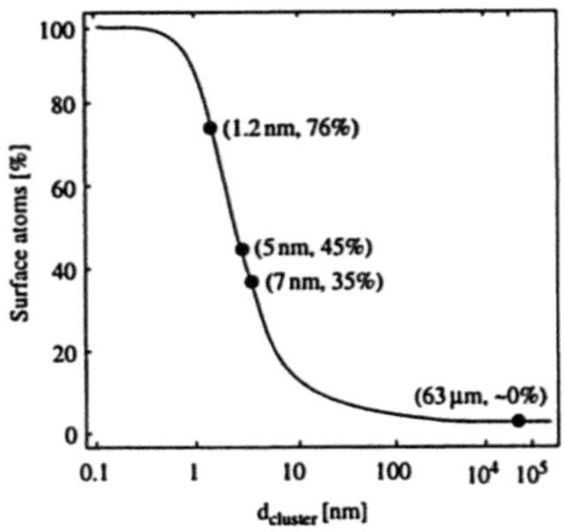

Figure 2.
The particle size reduction's effect on the surface atoms ratio [4].

a material are ultimately determined by this quantization of energy, also known as quantum confinement. In insulating materials like transition metal oxides, imperfections govern a material's behavior; this impact is accentuated in nanoscale structures due to their large surface and interfacial areas [7].

Producing items with nanometer-sized dimensions is known as nano-fabrication, and nano-fabrication techniques are crucial for developing innovative nanoscale structures, electronics, and materials with distinctive features [8].

4. Application of nano-fabrication in medicine and industry

Nanotechnology has numerous industrial applications, including electronics, energy, medicine, and materials science. Some of the vital industrial applications of nanotechnology are:

Electronics: Smaller and more effective electrical devices, such as transistors, sensors, and memory components, are being created with the help of nanotechnology. Nanoscale materials with specialized electrical properties, such as carbon nanotubes and graphene, are well suited for use in electronic devices.

Energy: Batteries, fuel cells, and solar cells are among the energy storage and conversion technologies being improved with nanotechnology. Nanomaterials like quantum dots and nanowires are being used to enhance the performance of these devices.

Medicine: Nanotechnology is creating new medicine delivery systems and diagnostic devices. Diseases can be treated more precisely and successfully with nanoparticles since they can be designed to target particular cells or tissues in the body.

Materials Science: Nanotechnology is used to develop stronger, lighter, and more durable materials. Nanomaterials like carbon nanotubes and graphene have exceptional mechanical properties that make them ideal for aerospace and automotive applications. Overall, the industrial applications of nanotechnology are vast and varied, and they can potentially revolutionize many industries in the coming years.

The topic of "soft nanotechnology," based on the self-assembly of giant organic molecules like polymers and proteins, has a wide range of practical uses.

In the context of a circular economy, employing affordable and environmentally friendly bio-based materials to create organic nanostructures is particularly

Figure 3.
Nanofabrication: Techniques and industrial applications [9].

appealing. Targeted medicine delivery, quick testing, high-throughput gene sequencing, and recent vaccinations utilize nanoscale bio-molecular (**Figure 3**) [9].

Richard Zsigmondy, an Austro-Hungarian colloid chemist who won the Nobel Prize in Chemistry in 1925, pioneered tiny nanoparticle research. He developed the word 'nanometer' to describe particles such as gold colloids that he observed under a microscope [9].

Nanotechnology is the manufacture of objects atom by atom. The introduction of nanotechnology into production has been likened to the introduction of prior technologies that have significantly impacted modern society, such as plastics, semiconductors, and even electricity. Nanotechnology applications offer extreme material performance and lifespan increases for electronics, medicine, energy, building, machine tools, agriculture, transportation, and apparel.

Nanofabricated products are constantly being developed. The industrial sectors that target these products are electronics and semiconductors, computing and information technology, communications, defense, automotive, chemical, and medical industries.

Some of the most common products obtained from nano-fabrication include [7]:

- Semiconductors
- Nanowires
- Nanostructured particles
- Nanotubes
- Coatings, paints, and thin layers
- Nanoparticles
- Nano/microfluidic systems
- Integrated optics
- Microelectromechanical systems (MEMS)

- Defense, security, and protection equipment
- Telecommunication products, displays, and optoelectronics

Chemists continue to explore the potential of nanotechnology and work toward developing new applications and materials that can further improve our lives.

5. Nano-fabrication Laboratory

The Nano-fabrication Laboratory (Nanofab) offers a range of state-of-the-art equipment and expertise in micro- and nano-fabrication, including photolithography, electron beam lithography, thin film deposition, etching, and characterization. The facility is equipped with cleanroom facilities essential for producing high-quality nanoscale structures. The Nanofab also provides training and support to researchers and students from various academic disciplines, including physics, chemistry, materials science, engineering, and biology. The facility is dedicated to fostering interdisciplinary research collaborations and promoting innovation in nanotechnology. The Nanofab is an essential resource for researchers who require access to advanced nanofabrication and characterization tools. By providing access to cutting-edge equipment and expertise, the facility enables researchers to explore new scientific frontiers and develop innovative technologies that have the potential to impact a wide range of fields, from medicine and electronics to energy and environmental science [10].

6. The difference between nano-fabrication and microfabrication

Nano-fabrication and microfabrication are two closely related fields of technology that involve manufacturing small-scale structures and devices. There is no one recognized definition of nano-fabrication or one that distinguishes it from microfabrication. As a result, what was formerly known as microfabrication has been renamed nano-fabrication, while the underlying concepts have remained substantially the same. Nano-fabrication refers to creating structures and devices on a nanoscale level, typically ranging from 1 to 100 nanometers in size. This involves using specialized tools and techniques, such as electron beam lithography, atomic layer deposition, and nanoimprint lithography, to manipulate materials at the atomic and molecular levels. Microfabrication, on the other hand, involves the creation of structures and devices on a larger scale, typically ranging from a few micrometers to a few millimeters in size. This field uses photolithography, etching, and deposition processes to create complex patterns and structures on various materials, including silicon, glass, and polymers. Both nano-fabrication and microfabrication have a wide range of applications in electronics, medicine, energy, and materials science. These technologies have enabled the development of smaller, faster, and more efficient devices and new materials with unique properties and functions [11].

Nano-fabrication is a rapidly advancing field with great promise for creating new and innovative materials and technologies that can revolutionize many industries. The most sophisticated manufacturing technique today is nano-fabrication, the future technology. This technology is sometimes known as "extreme technology" since it allows scientists to approach virtually the theoretical limit of precision, i.e., the size of a molecule or atom. Indeed, the manipulation of matter at the nanoscale

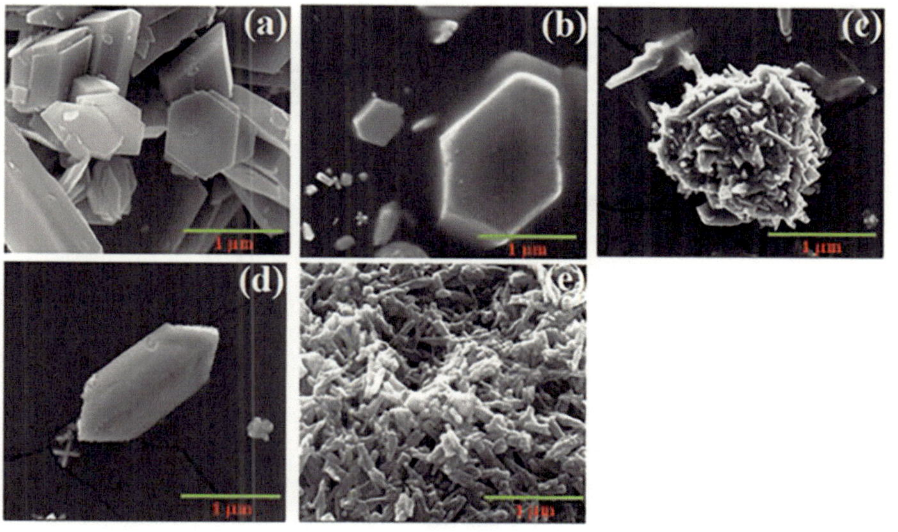

Figure 4.
SEM image of micro-nanoparticle [Ag(p-OH-C6H4CO2)2(NO3)]n [14].

may yield a wide range of materials and technologies that are significantly superior in performance, efficiency, and durability than those produced by traditional procedures. This nanoscale modification changes the material properties without impacting the substrate's fundamental qualities, making them fundamentally distinct and significantly superior to their bulk equivalents [12, 13] (**Figure 4**).

7. Classification of nano-fabrication

Nano-fabrication is a multidisciplinary science loosely defined as lithography, thin films, pattern transfer, and metrology. Another important subfield of "metrology" is inspection and characterization. We frequently create parts that are too small to be precisely examined with an optical microscope. There are various methods employed, including atomic force microscopy and electron microscopy. It may also be necessary to measure and keep track of additional characteristics, including film thickness, tension, and refractive index. Each of them calls for special equipment and methods [3].

8. Nanofabrication techniques

Nano-fabrication is a natural step toward further reducing the physical size of components and functional parts, and it frequently employs the same technology as microfabrication.

In summary, nano-fabrication is a critical aspect of nanotechnology research that involves the creation of structures and devices at the nanoscale. It requires a combination of "top-down" and "bottom-up" strategies and interdisciplinary collaboration between scientists from various fields. Some standard techniques include electron beam lithography, focused ion beam milling, nanoimprint lithography, and self-assembly. These techniques require specialized equipment and expertise, so many

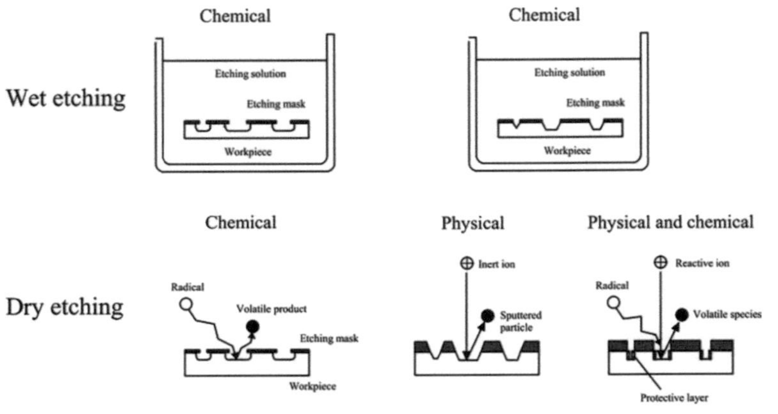

Figure 5.
Etching techniques [2].

researchers rely on nanoscience user facilities to conduct their experiments. Nano-fabrication techniques create structures such as nanowires, nanotubes, nanoparticles, nanocomposites, and devices such as sensors, actuators, and electronic components.

On the one hand, top-down procedures are used to create nanomaterials from a bulk substrate by eliminating material until the required nanomaterial is obtained; this category includes printing methods. Bottom-up methods, however, are precisely the reverse; the nanomaterial is generated starting at the atomic or molecular level and progressively assembling it until the required structure is formed [15]. As mentioned in previous discussions, nano-fabrication, on the other hand, deals with creating systems and devices at the nanoscale, typically defined as dimensions between 1 and 100 nm. At this scale, materials exhibit unique physical, chemical, and biological properties that can be harnessed for various applications. Nano-fabrication techniques create structures such as nanowires, nanotubes, nanoparticles, nanocomposites, and devices such as sensors, actuators, and electronic components. The methods used in nano-fabrication are similar to those used in microfabrication but with much higher precision and control. Some standard methods include electron beam lithography, focused ion beam milling, nanoimprint lithography, and self-assembly. These techniques require specialized equipment and expertise, so many researchers rely on nanoscience user facilities to conduct their experiments. In summary, nano-fabrication is a critical aspect of nanotechnology research that involves the creation of structures and devices at the nanoscale. It requires a combination of “top-down” and “bottom-up” strategies and interdisciplinary collaboration between scientists from various fields. We can unlock new possibilities for energy generation, electronics, medicine, and more by advancing our understanding of nano-fabrication [7, 16–18] (**Figure 5**).

9. Fabrication conditions

The conditions under which nano-fabrication occurs are crucial to the success of the process. Factors such as temperature, pressure, and impurities affect the final product. Additionally, the tools and equipment used in nano-fabrication must be carefully designed and maintained to ensure accuracy and precision. Cleanroom environments, where the air is filtered to remove particles and contaminants, are often used for nano-fabrication to minimize the risk of contamination.

The "nano-fabrication" requires dedicated facilities or laboratories, specialized equipment, fundamental skills, and, perhaps most crucially, a fascinating and multidisciplinary cultural environment to lead him or her through the everyday obstacles of discovering the "room at the bottom." A specialized climate is required to control each parameter during the various phases of nano-fabrication procedures. Given that materials' chemical and physical properties can vary greatly and that tools' thermal expansion might threaten the nanoscale scale's stability (and hence resolution), the temperature should be kept constant. Uncontrolled humidity may lead to the same issues. The same problems might arise from unchecked moisture. Regarding the proportions of manufactured items, dust particles may resemble a gigantic mountain.

When the nanoscale scientific research centers are merged, they serve as a gateway to additional major x-ray, neutron, and electron scattering user facilities. Each nanoscience user facility has clean rooms, nano-fabrication laboratories, one-of-a-kind signature equipment, and other tools (such as nanopatterning devices and research-grade probe microscopes). These facilities provide free access to cutting-edge instrumentation, computational methods, and expert scientific staff to researchers worldwide—a novel and standalone setting designed to encourage collaboration among scientists from various disciplines such as chemistry, biology, physics, materials science, engineering, and computer science [19, 20].

10. Nanoparticle types

Nanoparticles' size, shape, physical, and chemical features may be used to classify them. The categorization of nanoparticles is frequently determined by their function.

- **Carbon-based nanoparticles:** Carbon nanotubes (CNTs) and fullerenes are the two main types of carbon-based nanoparticles. These nanoparticles are extensively utilized in structural reinforcement since they are 100 times stronger than steel. Single-walled carbon nanotubes (SWCNTs) and multi-walled carbon nanotubes (MWCNTs) are the two varieties of CNTs. CNTs are unique in transferring heat down their length but not across the tube. Fullerenes are carbon allotropes with hollow cage structures composed of sixty or more carbon atoms. Buckminster fullerenes are the structure of C60, which resembles a hollow football.

- **Polymeric nanoparticles:** Polymeric nanoparticles are organic-based nanoparticles that are evenly disseminated. Polymeric nanoparticles have shapes that resemble noncapsular or nanospheres, depending on the technique of manufacture. A matrix-like structure characterizes nanosphere nanoparticles, whereas core-shell morphology characterizes nanocapsules. Active chemical and polymeric compounds are evenly spread in nanosphere polymeric nanoparticles, whereas a polymer shell in nanocapsule nanoparticles contain and encase the active compounds.

- **Dendrimers:** Dendrimers are a nanomaterial with a highly branched, tree-like structure. They are typically synthesized by repetitively adding layers of molecules to a central core. Dendrimers have several potential applications, including drug delivery, imaging, and catalysis. Their unique structure allows precise control over their size, shape, and surface properties, which can be tailored to specific applications. However, dendrimers can also be challenging to synthesize and purify, and their potential toxicity is still being studied.

- **Nanoshells:** Nanoshells are a type of nanomaterial that consist of a core material surrounded by a thin shell of a different material. They can be synthesized through various methods, including chemical vapor deposition and electrochemical deposition. Nanoshells have potential applications, including in biomedical imaging and cancer therapy. Their unique structure allows precise control over their optical properties, which can be tailored to specific applications. However, like dendrimers, nanoshells can also be challenging to synthesize and purify, and their potential toxicity is still being studied.

- **Nanowires**: Nanowires are another type of nanomaterial that consist of a long, thin wire-like structure with a diameter on the nanoscale. They can be made from various materials, including metals, semiconductors, and oxides. Nanowires have many potential applications, including electronics, energy storage, and sensors. Like nanoshells, their unique structure allows precise control over their properties, such as conductivity and surface area. However, their synthesis can also be challenging, and their potential toxicity is still being studied. Both nanoshells and nanowires are examples of the wide range of nanomaterials that are being developed and studied for their potential applications in various fields.

- **Lipid nanoparticles:** Lipid nanoparticles are a nanomaterial consisting of a lipid bilayer surrounding a core material, such as a drug or gene. They are commonly used in drug delivery applications, as the lipid bilayer can protect the cargo from degradation and increase its bioavailability. Lipid nanoparticles can be further classified into liposomes, solid lipid nanoparticles, and nanostructured lipid carriers, each with unique properties and applications. They are generally considered biocompatible and biodegradable, making them attractive for medical applications. However, temperature, pH, and lipid composition can influence their stability and release kinetics [21].

- **Ceramic nanoparticles:** Ceramic nanoparticles are a type of nanomaterial that consist of inorganic materials, such as oxides, nitrides, and carbides. They have unique physical and chemical properties, such as high hardness, melting point, and excellent thermal and electrical conductivity. Ceramic nanoparticles are commonly used in various applications, including catalysis, energy storage, and biomedical applications. In biomedical applications, ceramic nanoparticles have been used as drug-delivery vehicles, imaging agents, and bone substitutes. However, their potential toxicity and biocompatibility must be thoroughly evaluated before their use in medical applications. The properties of ceramic nanoparticles can be tailored by changing the particles' size, shape, and composition.

- **Metal nanoparticles:** Metal precursors are used to create metal nanoparticles, which can then be produced chemically, electrochemically, or photochemically.

- **Semiconductor nanoparticles:** Semiconductor nanoparticles, also known as quantum dots, are tiny particles made of semiconductor materials such as silicon, cadmium selenide, and zinc oxide. These particles are typically between 1 and 10 nanometers in size, making them much smaller than a human cell. Semiconductor nanoparticles have unique optical and electronic properties that make them useful in applications like Imaging, Solar cells, semiconductors, Lighting, Biomedical applications, and Environmental monitoring. Semiconductor nanoparticles can be used as fluorescent tags to label cells or tissues for imaging purposes. They emit light of a specific wavelength when excited by an external light source, allowing researchers to track their movement and behavior.

- **Hybrid nanoparticles** are particles made of two or more different materials, such as semiconductors and metal combinations. These nanoparticles have unique properties that differ from their components and can be tailored for specific applications. Some examples of hybrid nanoparticles and their applications are plasmonic, magnetic, and carbon-based nanoparticles. Overall, hybrid nanoparticles offer many possibilities for developing new materials with unique application properties.

Hybrid nanoparticles can be created through chemical processes such as hydrothermal, sol-gel, co-precipitation, photochemical, sonochemical, and seeding growth. In hydrothermal synthesis, the nanoparticles are formed in a high-pressure and high-temperature aqueous solution. Sol-gel synthesis involves the conversion of a sol into a gel by adding a cross-linking agent. Co-precipitation involves the simultaneous precipitation of two or more metal ions to form nanoparticles. In photochemical synthesis, nanoparticles are synthesized using light as an energy source. Sonochemical synthesis involves the use of ultrasound to create cavitation bubbles that lead to the formation of nanoparticles. Seeding growth consists of the use of pre-formed nanoparticles as seeds for the development of larger nanoparticles.

Physical approaches for creating hybrid nanoparticles include laser-induced heating, atom beam co-sputtering, and ion implantation. Laser-induced heating involves using a laser to heat a target material, causing it to vaporize and form nanoparticles. Atom beam co-sputtering involves simultaneously depositing two or more materials onto a substrate to form hybrid nanoparticles. Ion implantation consists of introducing ions into a solid material to create defects that can be used as nucleation sites for the formation of nanoparticles.

In addition to these methods, there are biological approaches for creating hybrid nanoparticles, such as using genetically modified organisms or biomolecules to synthesize nanoparticles. These methods are still in the early stages of development but can potentially be highly selective and environmentally friendly. The choice of fabrication method will depend on the desired properties and applications of the hybrid nanoparticles [22].

In the topic of examining nano-fabrication methods, we will get to know more about nano-fabrication techniques.

- **Thin film:** A thin film is a layer of material that has a thickness of a few nanometers to a few micrometers. Thin films can be made from various materials, including metals, semiconductors, and polymers. They have unique properties that differ from their bulk counterparts, such as higher surface area and improved optical, electrical, and mechanical properties. Thin films are commonly used in various applications, including electronics, optics, and energy storage. In electronics, thin films create transistors, sensors, and displays. In optics, thin films are used to develop anti-reflection coatings and filters. In energy storage, thin films make batteries and solar cells. The properties of thin films can be controlled by adjusting the deposition process, such as the material's temperature, pressure, and composition. With thicknesses on the order of a few nanometers, nanoscale thin films have unique characteristics that set them apart from bulk materials [23].

This conformal nature of thin films makes them ideal for coating complex three-dimensional objects, as they can fully cover all surfaces and maintain their properties uniformly. This property is advantageous in the electronics industry, where thin films create microelectronic devices on various substrates. The ability of thin films to conform to any surface also makes them useful in biomedical applications, such as drug delivery and tissue engineering.

Another critical aspect of thin film science is the ability to precisely control the thickness of the film. This is achieved through various deposition techniques, such as physical vapor deposition (PVD), chemical vapor deposition (CVD), and atomic layer deposition (ALD). Precise control over the thickness of a film is crucial for achieving desired properties, such as optical transparency or electrical conductivity. Thin film science has also played a significant role in the development of nanotechnology. These properties are exploited in various applications, such as nanoelectronics, nanophotonics, and nanobiotechnology. Thin film science has made it possible to precisely control the thickness and composition of these nanoscale layers. It allows the creation of new materials with desired functional properties, such as the nature of layer compliance, film thickness, dielectric constant, stress, chemical composition, and electrical conductivity [3].

In conclusion, thin film science is a multidisciplinary field with numerous applications in various industries. The conformal nature and precise thickness control of thin films make them ideal for coating complex objects and creating novel materials with tailored properties. With the continued advancement of thin film science, we can expect to see even more innovative applications.

Physical methods include techniques such as PVD and sputtering, where atoms or molecules are ejected from a solid source and deposited onto a substrate. Chemical processes include techniques such as CVD and ALD, where a chemical reaction occurs between precursor gases and the substrate surface, resulting in the deposition of a thin film.

Each method has its advantages and disadvantages, and the choice of method depends on factors such as the desired properties of the film, the substrate material, and the scale of production. For example, PVD is a widely used technique for depositing thin metallic films with high purity and uniformity, while CVD is often used for depositing thin films of oxides and nitrides with precise stoichiometry [24].

In recent years, hybrid methods that combine physical and chemical processes have also emerged, such as plasma-enhanced CVD and atomic layer etching. These methods offer greater control over the deposition process and can result in thin films with unique properties [22].

Overall, thin film science continues evolving and offers new possibilities for creating advanced materials with tailored properties. As technology advances, we can expect to see even more innovative applications of thin films in fields such as electronics, optics, energy, and biomedicine.

With the advance of nano-fabrication methods, manufacturing and functional synthesis of nanoparticles for medicinal, energy generation, and chemical engineering sectors will continue.

11. Classification of nano-fabrication methods

11.1 Chemical and physical nano-fabrication methods

11.1.1 Chemical methods

Chemical methods are an essential subset of nano-fabrication techniques that involve chemical reactions to create or modify nanoscale structures. Some standard chemical methods used in nano-fabrication include:

1. Sol-gel synthesis: This method involves forming a gel-like material from a solution of precursor molecules, followed by drying and heating to create a solid material with nanoscale features.

2. Electrochemical deposition: This method uses an electric current to deposit metal ions onto a substrate, creating a thin film or patterned structure.

3. Chemical vapor deposition: This method involves the reaction of gases to deposit a thin film of material onto a substrate. It is commonly used in semiconductors to create thin films of silicon and other materials.

4. Bottom-up synthesis: This method involves the assembly of individual atoms or molecules into larger structures using chemical reactions. It is often used to create nanoparticles and nanowires with specific properties.

5. Self-assembly: Self-assembly is a chemical method involving the spontaneous organization of molecules or nanoparticles into ordered structures.

Chemical methods are beneficial for creating complex, three-dimensional structures with precise control over their composition and properties. They are also crucial for creating functional materials with unique properties, such as catalytic nanoparticles or biomimetic materials. The most prevalent chemical approach is chemical vapor deposition (CVD). In this process, gas precursors are delivered into a chamber, and the substrate is heated to a sufficient enough temperature to initiate a reaction and form the film of interest. There are several forms of CVD, such as low-pressure CVD (LPCVD), microwaves CVD(MWCVD), radio frequency (RFCVD), metal-organic precursors CVD (MOCVD), realized in fluidized bed (fluidized bed CVD), UV beam CVD (photo CVD), atmospheric pressure CVD (APCVD) plasma-enhanced CVD (PECVD), Chemical vapor infiltration (CVI) and atomic layer deposition (ALD) [25].

Each of these processes has its advantages and limitations, and the choice of operation depends on the specific application and desired properties of the resulting material. For example, MOCVD is commonly used in the semiconductor industry to deposit thin films of materials such as gallium arsenide, while LPCVD is often used for depositing high-quality silicon nitride films. CVD processes are widely used to fabricate various nanostructures, including nanowires, nanotubes, and thin films. These structures have electronics, photonics, catalysis, and energy storage applications. For example, due to their unique electrical and mechanical properties, carbon nanotubes are being investigated for their potential use in next-generation electronics. Overall, CVD processes offer a versatile and scalable method for synthesizing nanostructures with precise control over their size, shape, and composition. As research in nanotechnology advances, we can expect to see even more innovative CVD processes and applications [26].

11.1.2 Physical methods

Physical methods are also widely used in nano-fabrication to create or modify nanoscale structures. Some of the most commonly used physical methods in nano-fabrication include lithography, etching, deposition, and imaging.

1. Lithography uses a patterned mask to selectively expose or remove material from a substrate, creating a patterned structure. This method is widely used in semiconductors to create microchips and other electronic devices.

2. Etching involves selectively removing material from a substrate using chemical or physical means. This method is commonly used in the semiconductor industry to create patterns and structures on a substrate.

3. Deposition involves depositing material onto a substrate using physical means, such as sputtering or evaporation. This method is widely used in the semiconductor industry to create thin films of metals and other materials with precise control over their thickness and composition.

4. Imaging involves using microscopy and other techniques to visualize and manipulate nanoscale structures. This method is widely used in research and development to study nanomaterials' properties and develop new nano-fabrication techniques.

5. Epitaxy is a physical method used in nano-fabrication to grow thin film materials on a substrate with controlled crystal orientation and thickness. This method involves depositing atoms or molecules onto a substrate in a specific pattern, then self-assemble into a crystal structure.

Epitaxy is widely used in semiconductors to produce high-quality, single-crystal films for electronic devices such as transistors and solar cells. It is also used in materials science research to study the properties of thin films and their interfaces. Advanced epitaxy techniques have enabled the growth of complex multilayer structures and heterostructures with tailored properties for specific applications. Ongoing research in epitaxy is focused on developing new materials and optimizing growth conditions to improve the performance of electronic and optoelectronic devices [3, 27, 28].

There are two main types of epitaxy: chemical vapor deposition (CVD) and molecular beam epitaxy (MBE). In CVD, the material is deposited onto the substrate in a gas phase, while in MBE, the material is deposited in a vacuum using a beam of atoms or molecules. Epitaxy is widely used in semiconductors to create high-quality thin films for electronic devices such as transistors and solar cells. It also produces LEDs, lasers, and other optoelectronic devices. Overall, epitaxy is an important technique in nano-fabrication that enables the precise control of crystal orientation and thickness, allowing for the creation of high-performance electronic and optoelectronic devices.

Physical vapor deposition (PVD) is another commonly used method for depositing thin films onto substrates. Unlike CVD, PVD involves the physical evaporation of a solid material, which then condenses onto the substrate to form a thin film. This can be achieved through various techniques, such as sputtering or thermal evaporation. One advantage of PVD is that it can produce high-quality films with excellent adhesion and uniformity. It also allows for precise control over the thickness and composition of the deposited film. Additionally, PVD can deposit various materials, including metals, alloys, and ceramics [25, 23].

However, PVD has some limitations. For example, it may not be suitable for depositing certain materials, such as polymers or organic compounds. It also requires high vacuum conditions, limiting its scalability for large-scale production. Despite these limitations, PVD is widely used in various industries, including electronics, optics, and aerospace. It is beneficial for producing coatings with specific properties, such as corrosion or wear resistance. As with CVD, ongoing research is focused on developing new PVD techniques and applications for nanotechnology. Overall, physical methods are important tools in nano-fabrication, enabling the creation of precise patterns and structures with high resolution and accuracy [2].

There are several methods for nano-fabrication, including:

1. Lithography: This technique uses light or electrons to create patterns on a substrate. It is commonly used in the semiconductor industry to make microchips.

2. Self-assembly: This method spontaneously organizes molecules or nanoparticles into ordered structures. It is often used to create nanoscale patterns or coatings.

3. Deposition: This method involves depositing thin films of material onto a substrate using techniques such as sputtering, evaporation, or chemical vapor deposition.

4. Etching: This method involves selectively removing material from a substrate using chemical or physical processes. It is often used to create patterns or structures in a material.

5. Nanoprinting: This method involves using specialized printers to deposit nanoscale materials onto a substrate. It is often used in the fabrication of sensors and other electronic devices.

Nano-fabrication methods are critical for developing nanotechnology and creating new materials and devices with unique properties and applications [3].

11.2 Top-down and bottom-up nano-fabrication methods

Bottom-up methods offer greater control over the resulting nanostructures' shape, size, and composition but are often limited in scalability and reproducibility. Top-down approaches, on the other hand, can be used to create large numbers of identical structures with high precision but may suffer from issues such as surface roughness and damage. Nano-fabrication—the design and production of structures at nanoscale dimensions—requires advances in material synthesis, physical patterning, structural characterization, and theory, including fields like biology, chemistry, physics, and engineering. As a result, researchers from all around the world are working to develop "top-down" and "bottom-up" methods for designing surface nanostructures [16].

"Top-down" methods directly imprint a design into a substrate using sophisticated lithography, electron-beam writing, and nanoimprinting methods. Due to their historical roots, these methods are particularly successful in creating nanostructures based on metal and semiconductors. "Bottom-up" techniques use atoms and molecules that spontaneously coalesce into well-organized structures. Nature frequently uses self-assembly.

We may combine nanosystems and create materials specifically suited for a specific purpose by combining "bottom-up" and "top-down" approaches. To understand a material's size-dependent qualities in a particular application while improving molecule and material design, extra care must be given when a material's dimensions decrease [17].

For instance, the bandgap of a nanoscale material determines optical characteristics, whereas exchange interactions between spin states control magnetic properties [7]. It is expected that both "top-down" and "bottom-up" approaches will need to collaborate with larger-scale elements in functional devices to successfully manufacture nanostructures with resolutions below 10 nm [2]. To uncover structure-property correlations, this calls for an atomic-scaled examination of nanomaterials.

The development of atomic-scale imaging methods like scanning tunneling microscopy and atomic force microscopy, as well as developments in classical electron microscopy, has substantially benefited attempts to visualize nanoscale structures and their alteration under external disruption [18].

Top-down nano-fabrication methods involve using larger structures or materials to create smaller designs or devices. These methods are typically used in the semiconductor industry, including photolithography, electron beam lithography, and focused ion beam milling. Photolithography is a process that uses light to transfer a pattern onto a substrate coated with a photosensitive material. This method is commonly used to create patterns on silicon wafers to produce microprocessors and other electronic devices. Overall, top-down nano-fabrication methods are essential for building electronic devices and other nanotechnology applications, as they enable the creation of precise patterns and structures at the nanoscale.

11.2.1 Top-down methods

11.2.1.1 Lithography

Several major nano-fabrication methods are coming from laboratories or are currently available on the market, including electron. Soft lithography, micro or nano-stencil-guided deposition, dip-pen or fountain-pen lithography, nano-xerography, scanning nanolithography, and nanoimprint lithography are all examples of lithography techniques [3].

Lithography is a physical method that can be the main idea of top-down methods derived from the nano-fabrication methods for creating hybrid nanoparticles, which involves using a patterned mask to selectively deposit or remove material from a substrate. This method can be used to create complex patterns and structures at the nanoscale and is commonly used in the semiconductor industry to fabricate microchips. There are several types of lithography, including photolithography, electron beam lithography, and nanoimprint lithography. Photolithography uses light to transfer a pattern from a mask onto a photosensitive material, while electron beam lithography uses a focused beam of electrons to directly write patterns onto a substrate. Nanoimprint lithography involves pressing a patterned stamp onto a substrate

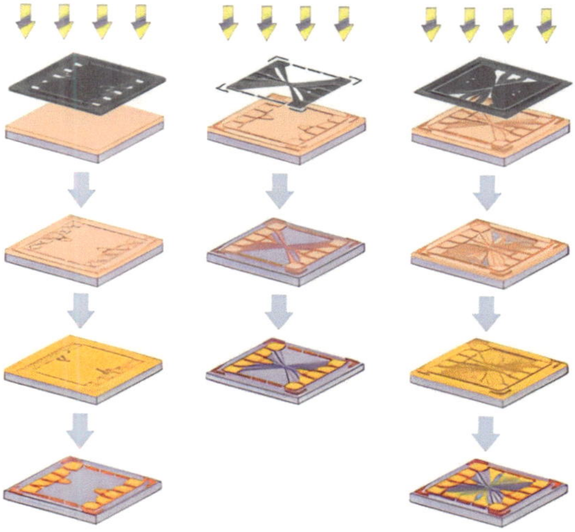

Figure 6.
Basic lithography process. (ET 1039—Nanotechnology, Alejandro Soliva Beser. Nanotechnology Fabrication Methods).

to transfer the pattern. Lithography is an exact and versatile method for creating hybrid nanoparticles, but it can also be time-consuming and expensive. It is typically used for small-scale production or research purposes rather than large-scale industrial applications. In the following, we examine examples of this method. Each method has its advantages and disadvantages [8] (**Figure 6**).

11.2.1.2 Conventional lithography

Conventional lithography is a top-down nano-fabrication method that uses a mask and a light source to pattern a substrate. The act contains a pattern transferred onto the substrate using a photoresist material. The mask exposes the photoresist to light, causing a chemical reaction allowing the pattern to be etched onto the substrate. Conventional lithography has been widely used in the semiconductor industry to create integrated circuits and other electronic devices. However, it has limitations regarding resolution and scalability, as the size of the features that can be patterned is limited by the wavelength of the light used. To overcome these limitations, alternative lithography techniques have been developed, such as electron beam lithography, which uses a focused beam of electrons to pattern the substrate, and nanoimprint lithography, which uses a stamp to transfer a pattern onto the substrate.

A traditional lithography process for producing integrated circuits (ICs) exposes a resist to a powerful particle beam, such as electrons, photons, or ions, by passing a flood beam through a mask or scanning a focused beam. The particle beam changes the chemical structure of the exposed portion of the resist layer. In the subsequent etching, the exposed or unexposed area of the resist will be erased to reproduce the patterns [29].

11.2.1.3 Photolithography

Photolithography is a type of lithography that uses light to transfer a pattern from a mask onto photosensitive material. The process involves several steps:

1. **Cleaning**: The substrate is cleaned to remove contaminants that may interfere with the patterning process.

2. **Spin coating**: A thin layer of photoresist is spun onto the substrate, creating a uniform coating.

3. **Mask alignment**: The mask is aligned with the substrate using a precision alignment system.

4. **Exposure**: The substrate is exposed to light through the mask, which causes the photoresist to undergo a chemical reaction and become either more or less soluble in a developer solution.

5. **Development**: The substrate is immersed in a developer solution, which removes the exposed areas of the photoresist, leaving behind a patterned layer.

6. **Etching**: The patterned layer is used as a mask for etching or deposition processes to create the desired nanoparticle structure [30] (**Figure** 7).

11.2.1.4 Advanced lithography

Advanced lithography technologies have been created to circumvent traditional lithography limits. Extreme ultraviolet (EUV) lithography, which employs a 13.5 nm wavelength light source, and multiple patterning, which requires repeated exposures and etching operations to make tiny features, are two technologies. EUV lithography is now employed to produce modern microprocessors, while multiple patterning creates memory devices and other advanced integrated circuits. Nanoimprint lithography, which utilizes a stamp to generate patterns on a resist, and electron beam lithography, which employs a focused electron beam to make high-resolution patterns, are two further advanced lithography approaches. These improved lithography processes allow for the fabrication of smaller and more complex structures, which is critical for developing modern electrical devices.

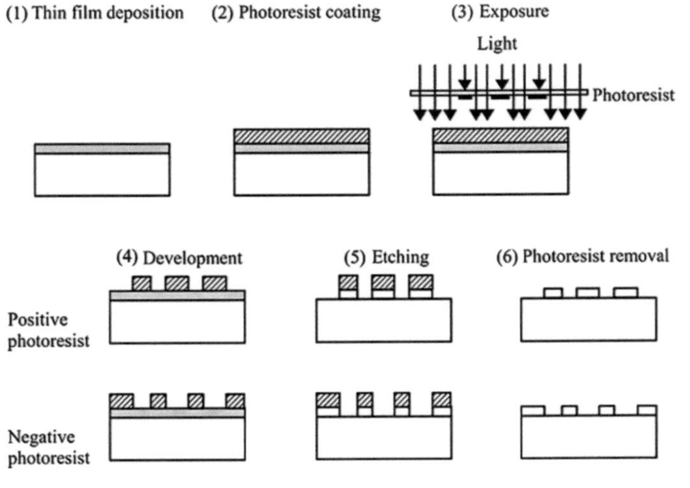

Figure 7.
Photolithography process [2].

DOI: http://dx.doi.org/10.5772/intechopen.112429

Another important area of research in advanced lithography is the development of new resists, which are materials used to transfer a pattern from a mask or template onto a substrate. Resists are critical in lithography, determining the resolution, contrast, and sensitivity. New resist materials are being developed to withstand higher temperatures and exposure doses, providing better pattern fidelity and lower line-edge roughness.

Overall, advanced lithography methods are essential for the continued progress of the semiconductor industry. As device dimensions continue to shrink and new materials and structures are developed, lithography will play an increasingly important role in enabling these advances [29].

11.2.1.5 Soft lithography

Soft lithography is a type of lithography that uses flexible materials, such as elastomers, to create patterns on substrates. This technique has many potential applications in the biotechnology, microfluidics, and optoelectronics fields. It is crucial to weigh soft lithography's potential benefits and risks and ensure it is used responsibly and sustainably. The lithographic process consists of coating a substrate with a resist, exposing the resist to light or electron beams, and developing the resist image with a chemical substance. The pattern is then transferred from the resist to the substrate through various techniques, such as chemical etching or dry plasma etching.

There are two main types of lithography: mask lithography, which uses a physical mask to irradiate the resist, and scanning lithography, which uses a scanning beam to irradiate the resist sequentially. Mask lithography is faster but has a lower resolution while scanning lithography is slower but has better resolution. Contact mode photolithography replicates the image on the mask as it is, while projection mode photolithography reduces the image using an optical system. Extreme UV or X-ray lithography can achieve higher resolution but requires expensive equipment. One of the significant motives for inventing soft lithography, for example, was to reduce the feature size and cost of microelectronic devices [31] (**Figure 8**).

11.2.1.6 Nanosphere lithography

Nanosphere lithography is a bottom-up nano-fabrication method that uses self-assembled nanospheres monolayers as a template to pattern a substrate. The nanospheres are deposited onto the substrate, forming a close-packed array, then coated with a thin layer of thin material. The nanospheres are then removed, leaving behind a patterned substrate. Nanosphere lithography has advantages in terms of scalability and resolution, as the size of the nanospheres can be controlled, and the pattern can be transferred onto a large substrate area. It is also a relatively simple and cost-effective method compared to conventional lithography.

Nanosphere lithography has been used in various applications, such as plasmonics, biosensors, and solar cells. It has also been combined with other techniques, such as chemical vapor and electrochemical deposition, to create more complex structures. Overall, nanosphere lithography is a promising technique for nano-fabrication and offers an alternative to conventional lithography for specific applications [33] (**Figure 9**).

11.2.1.7 Colloidal lithography

Colloidal lithography is another method used in nano-fabrication to achieve atom-by-atom precision. It involves using a monolayer of colloidal particles as a mask

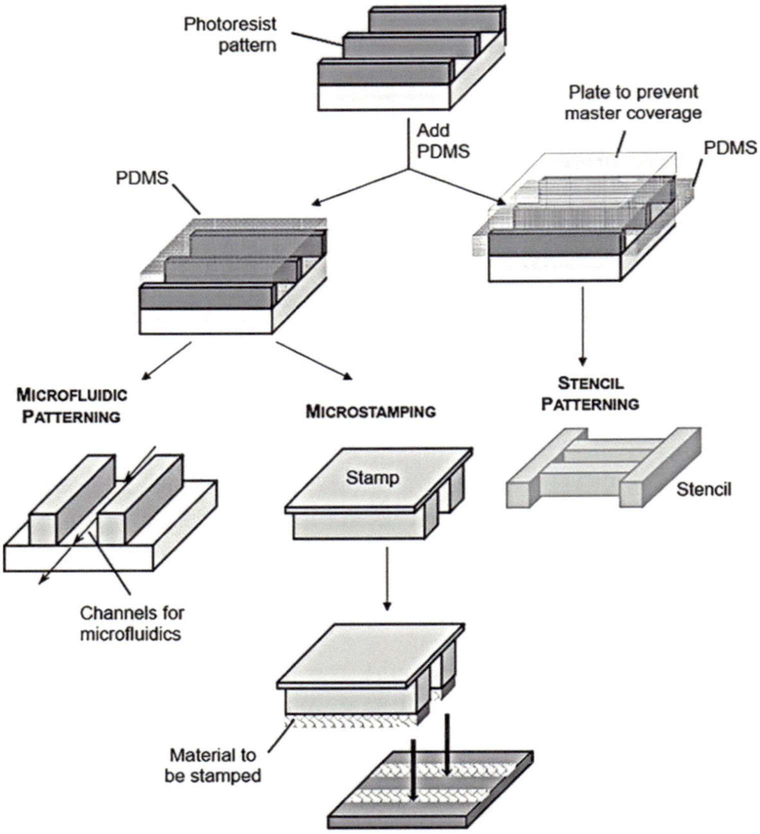

Figure 8.
Soft lithography (PDMS poly(dimethyl siloxane)) [32].

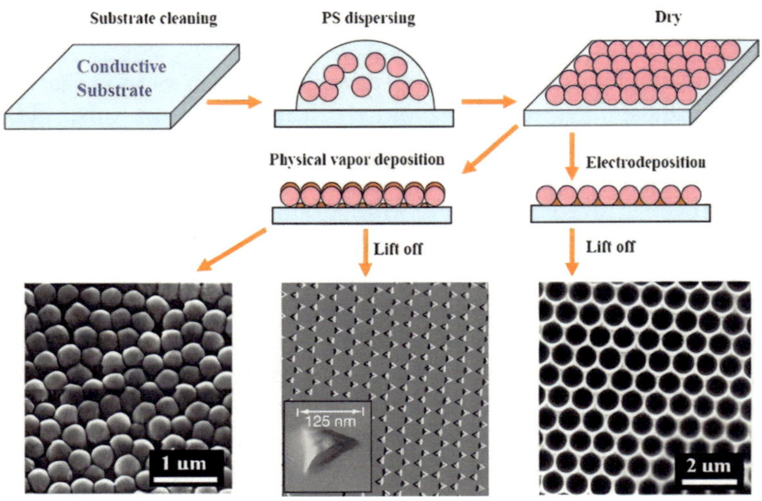

Figure 9.
Nanosphere lithography. (ET 1039—Nanotechnology, Alejandro Soliva Beser. Nanotechnology Fabrication Methods).

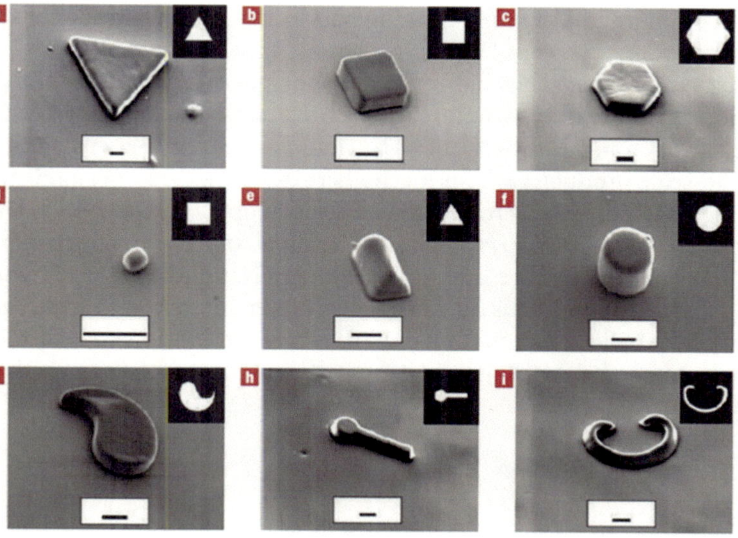

Figure 10.
Colloidal lithography. (ET 1039—Nanotechnology, Alejandro Soliva Beser. Nanotechnology Fabrication Methods).

to pattern a substrate. The substrate is then etched or coated with a material, and the colloidal particles are removed to reveal the desired pattern. This technique allows for the creation of complex patterns with high resolution and can be used to fabricate various nanostructures, including nanowires, nanodots, and nanorings. Colloidal lithography is similar to nanosphere lithography.

Colloidal lithography is a versatile and cost-effective method for achieving atom-by-atom precision in nano-fabrication. It is widely used in research and industry to fabricate electronic devices, sensors, and other nanoscale structures [34] (**Figure 10**).

11.2.1.8 Scanning probe lithography

Scanning probe lithography is a top-down method in nano-fabrication that uses a scanning probe microscope to manipulate and remove material from a surface at the

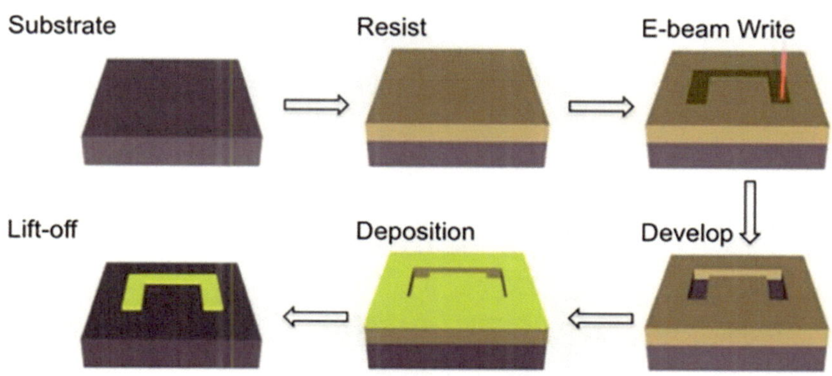

Figure 11.
Scanning probe lithography. (ET 1039—Nanotechnology Alejandro Soliva Beser. Nanotechnology Fabrication Methods).

nanoscale. This method allows for creating patterns and structures with high precision and resolution.

Scanning probe lithography has several advantages over traditional lithography methods, including creating structures on non-planar surfaces and working with a wide range of materials. However, it can be time-consuming, expensive, and unsuitable for large-scale production. This method creates structures mechanically moving a small or nanoscopic stylus across a surface to develop nanometer-scale patterns. SPL (Scanning Probe Lithography), which uses the tip of an AFM to selectively remove specific portions of the surface, and DPN (Dip-Pen Nanolithography), which uses an AFM tip to deposit materials on a surface with nanometer resolution, are popular. The primary benefits of these methods are their high resolution and capacity to make complex patterns with variable geometries, but their main restriction is their slow pace (**Figure 11**).

11.2.1.9 Writing "atom-by-atom"

Atom-by-atom precision is a key feature of nano-fabrication, as it allows for the creation of structures with minimal dimensions and high accuracy. By manipulating individual atoms and molecules, researchers can create materials with unique properties not found in bulk materials. One technique used to achieve atom-by-atom precision is scanning tunneling microscopy (STM), which allows researchers to image and manipulate individual atoms on a surface. By using a sharp tip to scan across the surface, STM can create patterns and structures with atomic-scale resolution (**Figure 12**).

Another technique is molecular beam epitaxy (MBE), which involves depositing atoms or molecules onto a substrate one layer at a time. Researchers can create thin films with precise thickness and composition by controlling the deposition rate and temperature [36, 37].

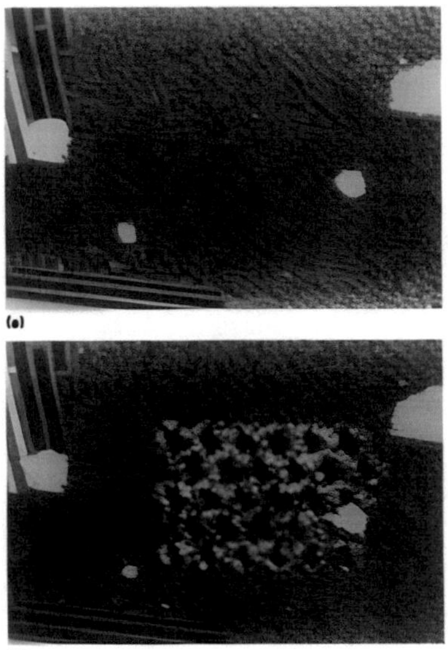

Figure 12.
The scanning tunneling microscope as a tool for nano-fabrication [35].

11.2.2 Bottom-up nano-fabrication methods

In contrast to top-down methods, bottom-up nano-fabrication methods involve the assembly of individual atoms or molecules to create larger structures or devices. Natural processes such as self-assembly and molecular recognition often inspire these methods. Self-assembly involves the spontaneous organization of molecules into a desired structure without external guidance. This method often creates nanoscale systems such as nanoparticles and nanowires. Molecular recognition involves the selective binding of molecules to each other based on their chemical properties. This method is often used to create functionalized surfaces and molecular sensors. Different bottom-up approaches include DNA nanotechnology, which uses DNA molecules as building blocks to create complex structures, and peptide-based nano-fabrication, which uses peptides to create functional materials and devices. Bottom-up nano-fabrication methods have the potential to enable the creation of new materials and devices with unique properties and functionalities and are an active area of research in the field of nanotechnology; in general, bottom-up methods offer a promising approach for the fabrication of nanomaterials, as they allow precise control over the size, shape, and composition of the resulting nanoparticles. As technology advances, bottom-up methods will likely become even more sophisticated, opening up new opportunities for nanotechnology applications. Bottom-up approaches are classified into two categories: gas-phase methods and liquid-phase methods.

In both situations, nanomaterials are created by a controlled manufacturing process that starts with a single atom or molecule and consists of small building blocks or molecules and atoms to create larger structures or substances. These methods rely on self-assembly, where building blocks are designed to interact with each other in a specific way to form the desired structure. Bottom-up processes offer advantages over top-down methods, such as creating complex systems with precise control over their properties and the potential for low-cost and scalable fabrication. However, these methods also have limitations, such as the difficulty in controlling the assembly process and the possibility of defects in the final product [37].

11.2.2.1 Plasma arching

Plasma arching is another technique used in nano-fabrication to synthesize and modify nanomaterials. It involves using plasma, a partially ionized gas, to generate high-energy species that can interact with and modify the surface of materials.

In plasma arching, a high-voltage electrical discharge directs plasma toward the substrate or material to be modified. The high-energy species in the plasma can cause chemical reactions on the material's surface, leading to the formation of new chemical bonds and the modification of its properties. Plasma arching can be used for a wide range of applications in nano-fabrication, such as surface cleaning, surface modification, deposition of thin films, and etching of materials. It offers several advantages over other techniques, such as high processing speed, precise control over the process parameters, and the ability to modify a wide range of materials.

However, plasma arching also has some limitations, such as the need for high-voltage equipment and potential damage to the processed material. Therefore, careful optimization of the process parameters is necessary to ensure this technique's successful and safe use in nano-fabrication. Plasma arcing is a highly efficient and scalable method for producing large quantities of nanotubes, making it a promising technology for future applications in electronics, energy storage, and other fields of cold plasmas; the average arc temperature is more than 104 K [38].

11.2.2.2 Chemical vapor deposition

Chemical vapor deposition (CVD) is a widely used method for synthesizing nanomaterials in nano-fabrication. In this process, a thin film of material is deposited on a substrate by the chemical reaction of gas-phase precursors. The process involves the following steps:

1. The substrate is placed in a reactor chamber and then evacuated to remove any air or moisture.

2. The precursors are introduced into the reactor chamber. These precursors can be in the form of gases, liquids, or solids that can be vaporized.

3. The precursors react with each other on the surface of the substrate, forming a thin film of the desired material.

4. The reaction conditions, such as temperature, pressure, and gas flow rate, are carefully controlled to ensure the desired properties of the thin film.

CVD can deposit various materials, including metals, semiconductors, and insulators. It is a versatile technique that can produce high-quality films with precise control over thickness, composition, and crystal structure. CVD is widely used in nano-fabrication to produce microelectronics, optoelectronics, and other advanced materials [39] (**Figure 13**).

11.2.2.3 Sol-gel synthesis

Another approach for producing thin films and nanoparticles is sol-gel synthesis. A precursor solution is created in this procedure by combining metal alkoxides or other inorganic chemicals with a solvent. After that, the solution is hydrolyzed and condensed to produce a gel, which may then be annealed or calcined to make a solid substance. The gel can be further processed by calcination or sintering to obtain the desired nanomaterials with controlled size, shape, and composition.

Sol-gel synthesis may create various materials, such as ceramics, glasses, and composites. The material's characteristics may be modified by varying the precursor composition, solvent, and production conditions.

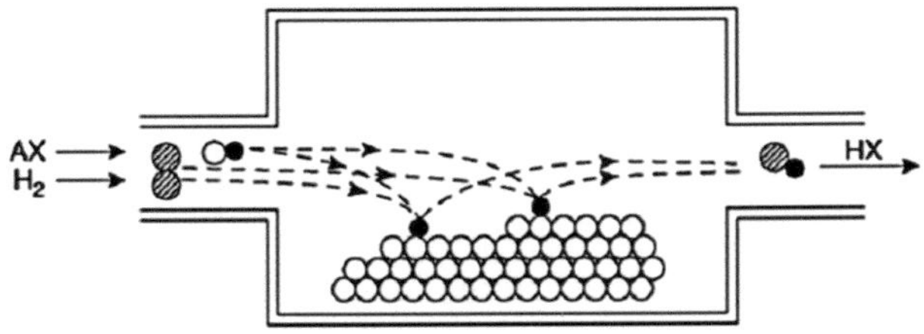

Figure 13.
The principle of CVD [39].

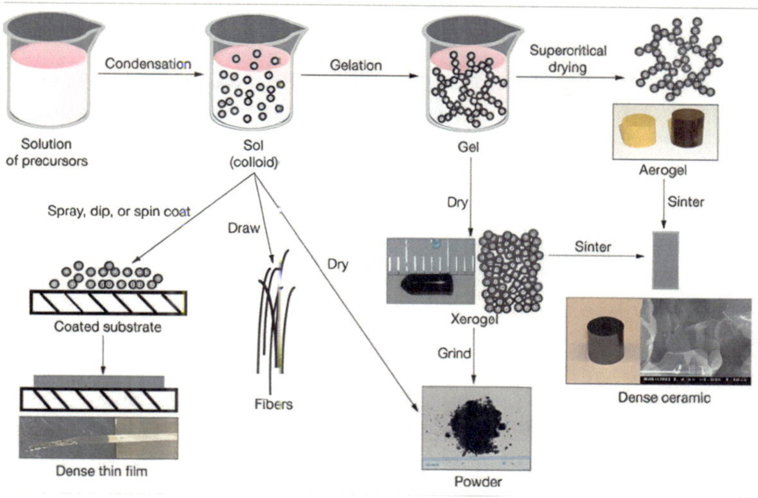

Figure 14.
Schematic overview of different materials obtained through a sol-gel process (image courtesy of Lawrence Livermore National Laboratory).

Sol-gel synthesis is widely used in nano-fabrication to produce catalysts, sensors, coatings, and other functional materials. Low processing temperatures, high purity, and the ability to construct diverse forms and architectures distinguish sol-gel synthesis from other nano-fabrication processes [40] (**Figure 14**).

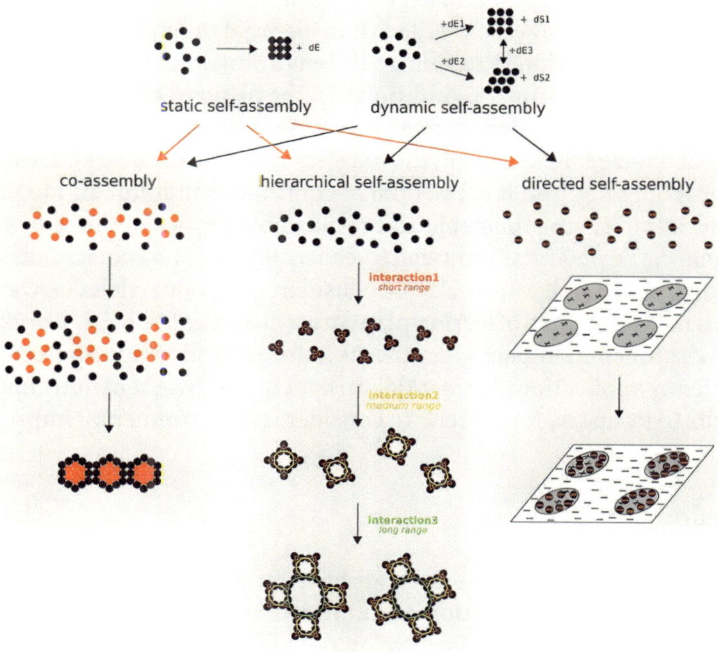

Figure 15.
Molecular self-assembly [41].

11.2.2.4 Molecular self-assembly

Molecular self-assembly is a bottom-up nano-fabrication method involving molecules' spontaneous organization into ordered structures without external manipulation. This process is driven by the inherent properties of the molecules, such as their shape, size, and interactions with each other and the substrate.

One example of molecular self-assembly is the formation of monolayers on a substrate, where molecules are adsorbed onto the surface and arranged in a specific pattern. This can create functional surfaces for various applications such as sensors, catalysis, and electronics.

Another example is the assembly of nanoparticles into larger structures through self-organization. This can create complex structures with unique properties for applications such as drug delivery, imaging, and energy conversion.

Molecular self-assembly offers several advantages over conventional top-down methods such as lithography. It is a scalable and cost-effective method that can create complex structures with high precision and control at the nanoscale. It also has the potential to develop new materials and devices with unique properties that cannot be achieved through conventional methods.

Overall, molecular self-assembly is a promising nano-fabrication method that can potentially revolutionize the field of nanotechnology [41] (**Figure 15**).

12. Final considering

When considering the industrial applications of nanotechnology, it is essential to consider the environmental impact of these technologies. Nano-fabrication methods often involve toxic chemicals and materials, negatively affecting the environment and human health if improperly handled. It is essential to implement sustainable practices in nano-fabrication to minimize the environmental impact and ensure the safety of workers and consumers. Additionally, the disposal of nanomaterials and products containing nanomaterials must be carefully managed to prevent potential environmental damage. Research on the long-term effects of nanomaterials on the environment is ongoing, and it is essential to continue to monitor and regulate the use of these materials to ensure their safe and responsible service in the industry.

Furthermore, the potential unintended consequences of nanotechnology on ecosystems and biodiversity must also be considered. Nanoparticles can accumulate in soil and water, potentially affecting plant growth and aquatic life. Conducting thorough environmental impact assessments before introducing nanotechnology into new industries or applications is essential. In conclusion, while nanotechnology offers many benefits to industry, it is crucial to consider its environmental impact [42].

13. Conclusion

In conclusion, nanofabrication methods have revolutionized various industries by enabling the precise manipulation and fabrication of materials at the nanoscale. These methods have allowed for the development of advanced technologies such as nanoelectronics, nanomedicine, and nanophotonics, which have significantly improved the efficiency, performance, and functionality of devices and systems. Nanofabrication methods, including top-down and bottom-up approaches, offer a wide range of techniques such as lithography, self-assembly, and deposition methods. These techniques provide control over the size, shape, composition, and arrangement of nanomaterials, leading to enhanced properties and novel functionalities. The use of nanofabrication methods has also led to the miniaturization of devices, allowing for the integration of more components on a single chip or substrate. This has resulted in smaller and more efficient electronic devices, sensors, and energy storage systems. Furthermore, nanofabrication methods have facilitated advancements in the field of nanomedicine, enabling the precise delivery of drugs and therapeutic agents to specific targets within the body. This has opened up new possibilities for personalized medicine and targeted therapies. However, there are still challenges and limitations associated with nanofabrication methods. These include high costs, scalability issues, and the need for specialized equipment and expertise. Additionally, the potential environmental and health risks associated with nanomaterials need to be carefully considered and addressed. Overall, nanofabrication methods have had a profound impact on various industries and hold great promise for future technological advancements. Continued research and development in this field will likely lead to further breakthroughs and applications in areas such as electronics, healthcare, energy, and environmental sustainability.

Author details

Saeedeh Ebad-Sichani[1], Paria Sharafi-Badr[2], Payam Hayati[3*]
and Mohammad Jaafar Soltanian-Fard[1]

1 Department of Chemistry, Firozabad Branch, Islamic Azad University, Firozabad, Fars, Iran

2 Department of Pharmacognosy and Pharmaceutical Biotechnology, School of Pharmacy, Iran University of Medical Sciences, Tehran, Iran

3 Organic and Nano Group (ONG), Department of Chemistry, Iran University of Science and Technology (IUST), Tehran, Iran

*Address all correspondence to: payamhayati@yahoo.com

References

[1] Fullerines MN, editor. Chapter 1 – Atom-by-Atom Manufacturing: The birth of nanotechnology. In: Fundamentals and Applications of Nano Silicon in Plasmonics. Elsevier; 25 June 2018. pp. 1-6. DOI: 10.1016/C2016-0-00112-2. ISBN 978-0-323-48057-4

[2] Kondoh E. Micro-and Nanofabrication for Beginners. Singapore: Jenny Stanford Publishing Pte. Ltd.; 2022. pp. 2-3. ISBN 978-981-4877-09-1

[3] Sarangan A. Nanofabrication: Principles to Laboratory Practice. London, USA: CRC Press; 2016. DOI: 10.1016/b978-1-78242-464-2.00005-1

[4] Cao G. Nanostructures & Nanomaterials: Synthesis, Properties & Applications. London: Imperial College Press; 2004. ISBN 1860944159 -1860944809

[5] Kasap SO. Electronic Materials and Devices. New York, NY: McGraw-Hill Education; 2006. ISBN 978-0-07-802818-2

[6] Sarkar D. Nanostructured Ceramics: Characterization and Analysis. Boca Raton, FL: CRC Press; 2018. DOI: 10.1201/9781315110790

[7] Scholes GD, Rumbles G. Excitons in nanoscale systems. Nature Materials. 2006;**5**(9):683-696

[8] Hu H, Kim HJ, Somnath S. Tip-based nanofabrication for scalable manufacturing. Micromachines. 2017;**8**(3):90

[9] Vasilev C. Nanofabrication: Techniques and industrial applications. In: AZoNano. Manchester, UK: AZoNetwork UK Ltd; 2021. Available from: https://www.azonano.com/article.aspx?ArticleID=5876

[10] Liddle JA, Gallatin GM. Nanomanufacturing: A perspective. ACS Nano. 2016;**10**(3):2995-3014

[11] Haus JW. Fundamentals and Applications of Nanophotonics. Woodhead Publishing; 2016. p. 426. ISBN: 9781782424642

[12] Wei D. Electrochemical Nanofabrication: Principles and Applications. Boca Raton, Florida, United States: Taylor & Francis, CRC Press; 2017. ISBN 9781315364919

[13] Zarnaghash N et al. Selective ultrasonic assisted synthesis of iron oxide mesoporous structures based on sulfonated melamine formaldehyde and survey of nanorod/sphere, sphere and core/shell on their catalysts properties for the Biginelli reaction. Materials Science and Engineering: C. 2019;**104**:109975

[14] Hayati P, Mehrabadi Z, Karimi M, Janczak J, Mohammadi K, Mahmoudi G, et al. Photocatalytic activity of new nanostructures of an Ag(i) metal–organic framework (Ag-MOF) for efficiently degrading MCPA and 2,4-D herbicides under sunlight irradiation. New Journal of Chemistry. 2021;**45**(7):3408-3417. DOI: 10.1039/d0nj02460k

[15] Yu H-D et al. Chemical routes to top-down nanofabrication. Chemical Society Reviews. 2013;**42**(14):6006-6018

[16] Barth JV, Costantini G, Kern K. Engineering atomic and molecular nanostructures at surfaces. Nature. 2005;**437**(7059):671-679

[17] Alivisatos P et al. From molecules to materials: Current trends and future directions. Advanced Materials. 1998;**10**(16):1297-1336

[18] Klabunde KJ, Richards RM. Nanoscale Materials in Chemistry. New York, NY: John Wiley & Sons; 2009. DOI: 10.1002/9780470523674

[19] Cabrini S, Kawata S. Nanofabrication Handbook. Boca Raton, Florida, United States: CRC Press; 2012

[20] Luttge R. Nano-and Microfabrication for Industrial and Biomedical Applications. William Andrew. Norwich, NY: ChemTec Publishing; 2016

[21] Singh NA. Nanotechnology innovations, industrial applications and patents. Environmental Chemistry Letters. 2017;**15**(2):185-191

[22] Tri PN et al. Methods for synthesis of hybrid nanoparticles. In: Noble Metal-Metal Oxide Hybrid Nanoparticles. Sawston, Cambridge: Elsevier, Woodhead Publishing; 2018. pp. 51-63. DOI: 10.1016/B978-0-12-814134-2.00003-6

[23] Hachem K et al. Methods of chemical synthesis in the synthesis of nanomaterial and nanoparticles by the chemical deposition method: A review. BioNanoScience. 2022;**12**(3):1032-1057

[24] Taschuk MT, Hawkeye MM, Brett M. Handbook of deposition technologies for films and coatings. Elsevier. 2010;**2010**:621-678

[25] Dobrzanski L, Staszuk D. Chemical vapor deposition in manufacturing. In: Nee A, editor. Handbook of Manufacturing Engineering and Technology. London: Springer; 2013

[26] Kern W, Schuegraf KK. 1 - deposition technologies and applications: Introduction and overview. In: Seshan K, editor. Handbook of Thin Film Deposition Processes and Techniques. Second ed. Norwich, NY: William Andrew Publishing; 2001. pp. 11-43

[27] Shchukin V, Ledentsov NN, Bimberg D. Epitaxy of Nanostructures. Springer Series on Nanoscience and Technology Berlin. Heidelberg: Springer Berlin; 2003. p. 450. DOI: 10.1007/978-3-662-07066-6_1

[28] Aliofkhazraei M, Ali N. PVD technology in fabrication of micro- and nanostructured coatings. Comprehensive Materials Processing. Amsterdam, Netherlands, USA: Elsevier; 2014. pp. 49-84. DOI: 10.1016/b978-0-08-096532-1.00705-6

[29] Yang L et al. Nano-fabrication: A review. Journal of the Chinese Institute of Engineers. 2007;**30**(3):441-446

[30] Stoykovich MP, Nealey PF. Block copolymers and conventional lithography. Materials Today. 2006;**9**(9):20-29

[31] Lipomi D et al. 7.11: Soft lithographic approaches to nanofabrication. Polymer Science Compression Reference. 2012;**10**:211-231

[32] Betancourt T, Brannon-Peppas L. Micro- and nano-fabrication methods in nanotechnological medical and pharmaceutical devices. International Journal of Nanomedicine. 2006;**1**(4):483-495. DOI: 10.2147/nano.2006.1.4.483

[33] Haynes CL, Van Duyne RP. Nanosphere lithography: A versatile nanofabrication tool for studies of size-dependent nanoparticle optics. The Journal of Physical Chemistry B. 2001;**105**(24):5599-5611. DOI: 10.1021/jp010657m

[34] Ademovic Z, Kingshott P. Micro-and nanoscale surface patterning techniques for localising biomolecules and cells: The essence of nanobiotechnology. In: Vadgama P editor. Surfaces and Interfaces for Biomaterials. Cambridge: Woodhead Publishing; 2005. pp. 150-180. DOI: 10.1533/9781845690809.1.150. ISBN (Print)1-85573-930-5

[35] van Loenen EJ, Dijkkamp D, Hoeven AJ, Lenssinck JM, Dieleman J. Direct writing on Si with a scanning tunneling microscope. Applied Physics Letters. 1989;**551**:312

[36] Ehrichs E, Silver R, De Lozanne A. Direct writing with the scanning tunneling microscope. Journal of Vacuum Science & Technology A: Vacuum, Surfaces, and Films. 1988;**6**(2):540-543

[37] Urban J et al. The scanning tunneling microscope as a tool for nanolithography: Writing nanostructures on Si (110) in air. Microelectronic Engineering. 1995;**27**(1-4):113-116

[38] Kumar S, Bhushan P, Bhattacharya S. Fabrication of nanostructures with bottom-up approach and their utility in diagnostics, therapeutics, and others. Environmental, Chemical and Medical Sensors. 2018:167-198

[39] Carlsson J-O, Martin PM. Chemical vapor deposition. In: Handbook of Deposition Technologies for Films and Coatings. William Andrew, Elsevier; 2010. pp. 314-363. DOI: 10.1016/b978-0-8155-2031-3.00007-7. ISBN 978-0-8155-2031-3

[40] Aguilar GV. Introductory chapter: A brief semblance of the sol-gel method in research. In: Sol-Gel Method-Design and Synthesis of New Materials with Interesting Physical, Chemical and Biological Properties. 2019:106 DOI: 10.5772/intechopen.76535. ISBN 978-1-78985-334-6

[41] Ozin GA et al. Nanofabrication by self-assembly. Materials Today. 2009;**12**(5):12-23. DOI: 10.1016/s1369-7021(09)70156-7

[42] Ray PC, Yu H, Fu PP. Toxicity and environmental risks of nanomaterials: Challenges and future needs. Journal of Environmental Science and Health, Part C. 2009;**27**(1):1-35. DOI: 10.1080/10590500802708267

Chapter 2

Theories, Hypothesis and Rules for Morphology Transition Engineering of 1D, 2D and 3D Nanomaterials

Ahmed Hashem Abdelmohsen, Sherif A. El-Khodary and Nahla Ismail

Abstract

Most of the chemical and physical properties of nanomaterials vary remarkably according to their size, shape, and structure. Thus morphology is a crucial parameter that controls the properties and functionality of materials. On the basis of Abdelmohsen et al.'s theories and hypothesis, which are theory for morphology transition engineering (ATMTE), theory for morphology engineering of solid compounds (ATMESC), and hypothesis for engineering of micro- and nanostructures (AHEMNS), novel approach was modified for fabricating one-, two-, and three-dimensional hybrid nanomaterials, such as hybrid ZnO nanosheets (38–150 nm), hybrid ZnO nanorods, hybrid nanocomposites, and hierarchical hybrid Cu_2O nanostructures. In addition, by the help of this novel method, the fabrication of metal-oxidene (one/few atoms thick layer of metal oxides) is assumed and hybrid ZnO thin film that is expected to have extraordinary physicochemical properties. A series of selection rules and morphology engineering rules are discussed. Throughout this chapter, we will come across this novel approach as a promising technique for nanofabrication and discuss the suggested mechanisms for the evolution process during fabrication of nanomaterials. By the help of this method, we have fabricated 1D, 2D and 3D nanomaterials that are expected to have potential use for energy, catalysis, biomedical, and other applications.

Keywords: Abdelmohsen et al. Theory 1 (ATMTE), Abdelmohsen et al. Theory 2 (ATMESC), Abdelmohsen et al. Hypothesis (AHEMNS), selection rules (SRs), morphology engineering rules, 1D, 2D and 3D nanomaterials, morphology transition, dangling bonds, morphology engineering, metal oxides, polyoxometalates (POMs), structural directing materials, Zn-oxidene, hybrid thin film

1. Introduction

Since the evolution of the field of nanotechnology "The field of research concerned with building up materials and devices on the scale of atoms and molecules" by the help of the pioneer work of R. Feynman, K. Drexler, and N. Taniguchi; the nanofabrication techniques evolved and developed enormously [1]. Nanomaterial is one-billionth of meter which is ten times the diameter of a hydrogen atom. More precise definition was postulated by B. Fahlman which state that "Nanotechnology is the creation of functional materials, devices and systems through control of matter on the nanometer length scale (1–100) nanometers and exploitation of novel phenomenon and properties (physical, chemical, biological, mechanical electrical, etc.) at that length scale" [1]. At such scales, the classical rules of chemistry and physics cannot be applied like melting point, reactivity, conductivity, strength and color can differ remarkably when the materials converted from bulk to nano-scale [2]. For example, in case of gold, there is a change in color from the bright yellow to red and a drop in melting point from 1336 K to 310 K when the size decreased [3]. There is a change in mechanical properties of carbon-based nanomaterials like fullerenes, nanotubes, and graphene. At nano-scale these materials become lighter and stronger than other structural materials like steel [4, 5]. Finally, other materials like ionic and metal carbon compounds experience changes in electrochemical and electronic properties at the nano-scale in terms of function or magnitude [6].

The concept of materials fabrication/engineering was expanded by Abdelmohsen et al. to include "controlling and designing the oriented structures of materials by re-scaling their dimensions or varying their external morphologies, favorably with functionalization, decoration, doping or mixing with other materials to attend the synergistic effect which enhance their properties" [7]. Nanofabrication techniques are classified into two main approaches which are top-down and bottom-up approaches [8]. The top-down approach involves the breaking down of the materials and converting it from bulk to nano-scale. On the other hand, bottom-up approach involves the building up of nanomaterials atom by atom and molecule by molecule. Scientists recommended the latter over the former approach, because it can fabricate structures with homogenous chemical composition, with better short- and long-range ordering, and less defects [8]. **Figure 1** illustrates an example for the bottom-up approach for engineering of nanomaterials like metal oxides atom by atom in solution. During this process the starting materials can be engineered to miscellaneous nanostructures like particles, rods, tubes, thin films, plates and stars [7]. **Figure 2** illustrates an example for the top-down approach for engineering of nanomaterials like carbon nanoparticles (CNPs). This process involves the breaking down of the bulk carbon black particles into smaller particles by ultra-sonication in the presence of polyoxometalates (POMs) like phosphomolybdic acid (PMA) which stabilizes the etched nanoparticles [7].

A precise definition for morphology transition engineering was postulated by Abdelmohsen et al. which is "The morphology evolution of materials under the effect of external stimulus (chemically, physically or mechanically) to build up new materials with various dimensions under easily attainable conditions (e.g., one step synthesis)" [9]. The role of the external stimulus is the inducing of the surface polarity by etching of the outer surface atoms to create reactive dangling bonds. The chemical stimulus is the acids which etch and dissolve the outer surface atoms to form the dangling bonds [7]. The physical stimulus is the accelerated particles that etch the outer surface atoms to form the dangling bonds [10, 11]. Finally, the mechanical stimulus involves the breaking down of the bulk particles or etching their outer atoms by ball milling to form also the dangling bonds [11, 12]. After the formation of reactive dangling bonds there are four possibilities that takes place which cause morphology transition engineering. The first possibility is the fusion of the etched particles to form different morphologies [7, 9]. The second possibility is the reconstruction

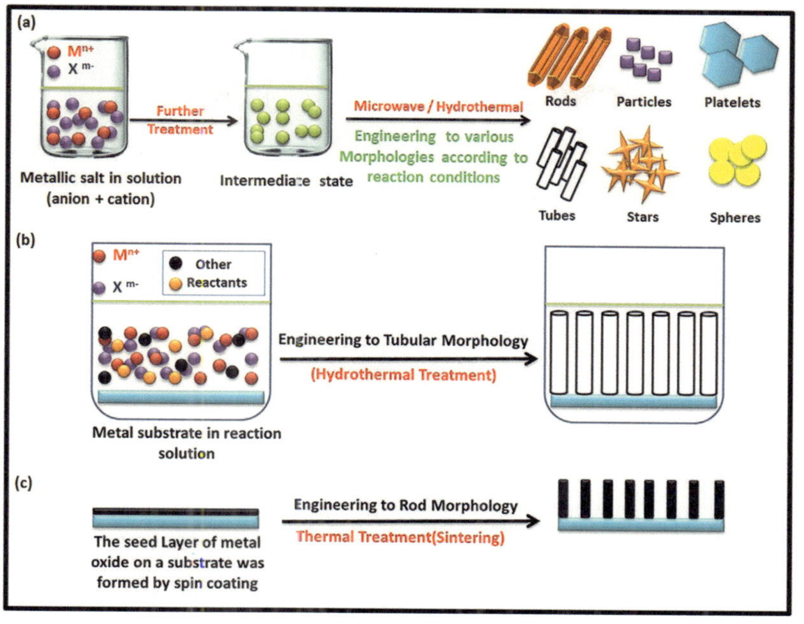

Figure 1.
Schematic illustration for the morphology engineering of metal oxides by bottom-up approach. Reproduced with permission [7] 2017, Springer Nature.

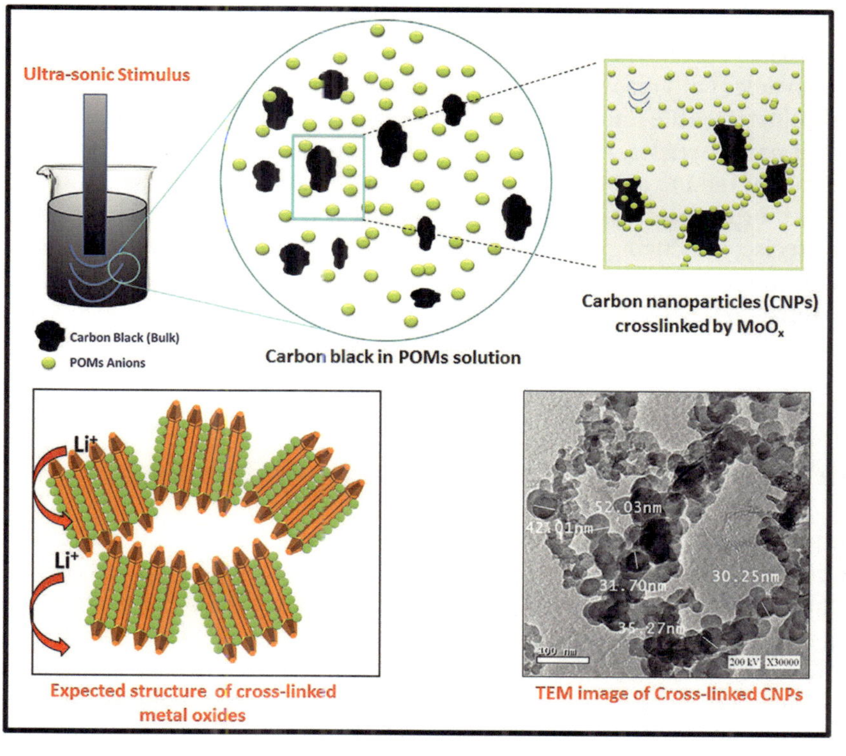

Figure 2.
Schematic illustration for the morphology engineering of carbon nanoparticles (CNPs) by top-down approach [Polyoxometalates assisted-solution technique]. Reproduced with permission [7] 2017, Springer Nature.

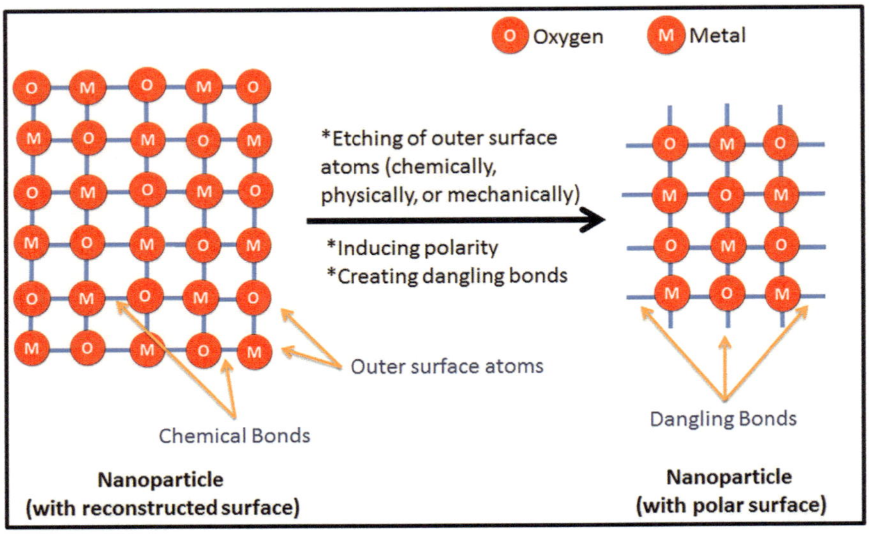

Figure 3.
Illustrates the chemical etching of MO nanoparticles to form smaller particles with reactive surfaces.

of the etched particles to form different morphologies [7, 9]. The third possibility is the adsorption of charged species to form also different morphologies by formation of atomic layers [11, 13]. The fourth possibility is the further growth of materials to different dimensions [7, 9]. The mechanism will be illustrated in details in Section 3. **Figure 3** shows the chemical etching of metal oxide (MO) nanoparticles to form smaller particles with reactive surfaces (polar surfaces/surfaces with dangling bonds). **Figure 4** illustrates the four possibilities that take place after etching/inducing surface polarity of nanoparticles. Reconstruction is a complementary process in all occasions as the particles tend to acquire the stable state after being in the active unstable state [7, 9]. The morphology transition process involves the top-down and bottom-up techniques. Precisely, inducing the surface polarity by etching the materials (chemically, physically, or mechanically) breaks down the bulk materials to smaller ones. This process refers to the top-down approach. The four possibilities mentioned take place to form further morphologies 1D, 2D and 3D nanostructures, refers to the bottom-up approach [7, 9].

2. Different studies on morphology transition engineering

The phase evolution of ZnO films that deposited electrochemically is greatly depending on the concentration of SO_4^{2-} ion as discovered by Wang and co-workers. As discovered by this group, the SO_4^{2-} ion concentration plays a critical role in engineering the morphology of the films from rod-like to plate-like structure. **Figure 5** shows the morphology evolution of ZnO from nanorods to nanoplatelets at different $ZnSO_4$ concentrations [14]. Das et al. studied the morphology evolution in hexagonal $V_{10}O_{28}$ – type polyoxometalate macrocrystals as a function of sonication temperatures. The morphology transformed from nano-rods to microflowers passing through intermediate hexagonal shaped microcrystals by raising the sonication temperature from 50°C to 80°C [15].

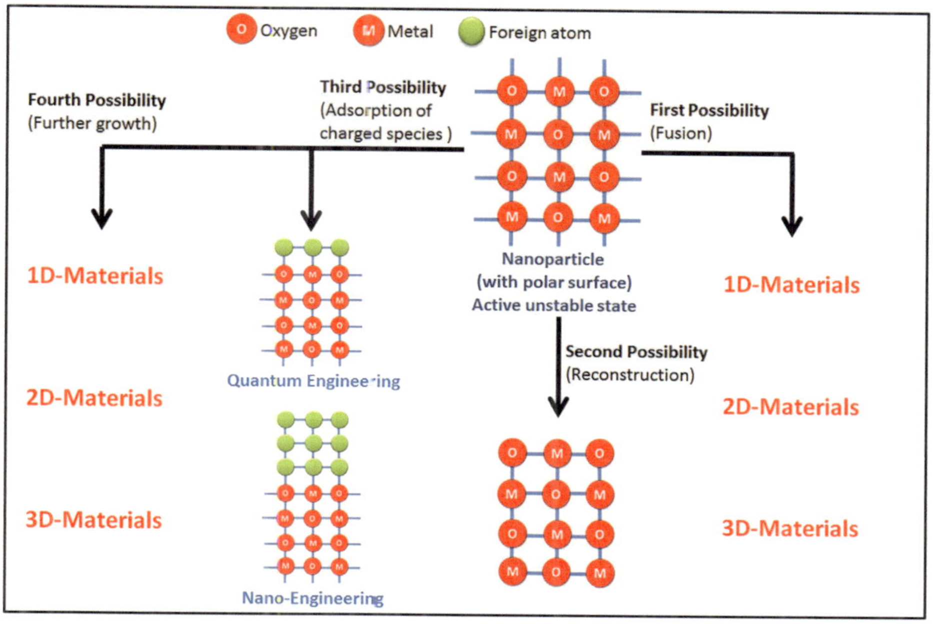

Figure 4.
Illustrates the four possibilities that take place after etching/inducing surface polarity of nanoparticles.

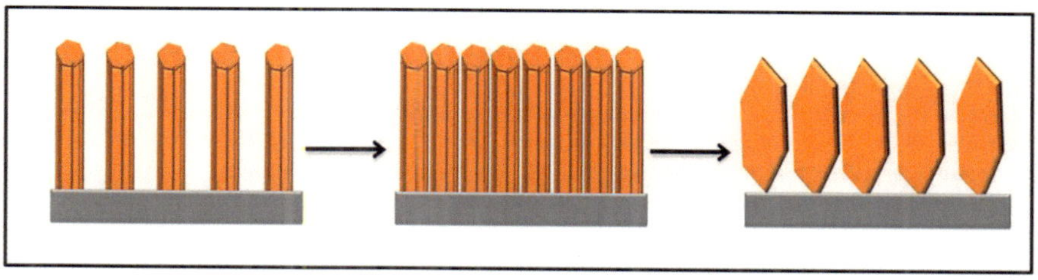

Figure 5.
Shows the morphology evolution of ZnO from nanorods to nanoplates by increasing concentration from left to right.

Xiong et al. used a facile binary-solution route and sequential thermal decomposition at atmospheric pressure to synthesize mesoporous 2D and 3D architectures of the oxide semiconductor Co_3O_4, including nanosheets, monodisperse microspheres are self-assembled from nanosheets, and copper-coin-like nanosheets. The volume ratio of H_2O and ethanolamine (EA) play a crucial role in the morphology of the precursor [16]. Yan et al. has reported the morphology evolution of single-crystalline octahedron CeO_2 to multi-nanostructure and nanorods using $Ce(NO_3)_3{\bullet}6H_2O$ as cerium resource and $Na_3PO_4{\bullet}6H_2O$ as mineralizer, the morphology evolution between the nano-octahedron and nanorod was observed by tuning the hydrothermal treatment time as shown in **Figure 6a** [17]. The Morphology Evolution of ZnO Thin Films from aqueous solutions was demonstrated by Goa and co-workers. A two-step approach was employed for the epitaxial growth of ZnO by beginning with the preparation of a (002)-oriented ZnO seed layer electrochemically. The substrate was soaked in an aqueous solution containing $ZnCl_2$ and complex agents. A large-scale fabrication of ZnO nanorods arrays on transparent conductive oxides has been

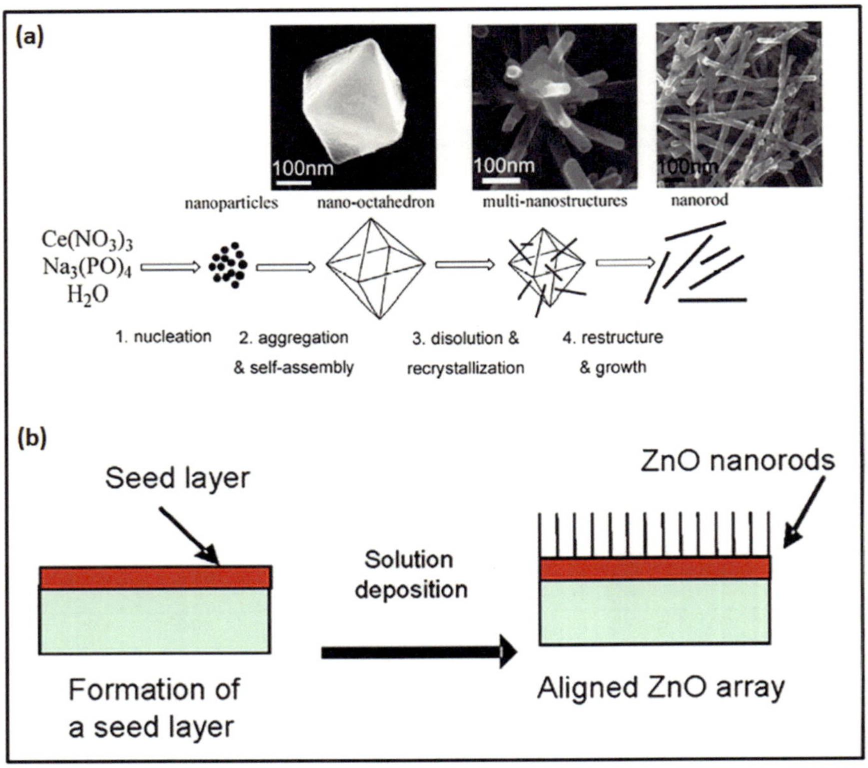

Figure 6.
Shows (a) schematic illustration for the multinanostructures evolution of CeO_2, (b) the morphology evolution of ZnO nanorods of ZnO thin films from aqueous solutions. Reproduced with permission [17] 2008, American Chemical Society. Reproduced with permission [18] 2006, American Chemical Society.

achieved after soaking at 95°C for 1–48 h as shown in **Figure 6b** [18]. Sweegers et al. discovered that the reaction conditions have a great effect on the formed morphology of gibbsite, g-$Al(OH)_3$. While the process of growth the gibbsite crystals morphology changed from thin, rounded hexagons and faceted lozenges into plates and blocks with well-formed basal, prismatic faces. The alkali ions of solution play the major role to control the morphology of gibbsite. An evidence for this fact is the engineering of different morphologies from potassium and cesium hydroxide solutions [19]. Morphology evolution of urchin-like $NiCo_2O_4$ nanostructures from urea was reported by Wang and co-workers. A'rods-to-straw-bundles-to-urchins' mechanism was proposed. Images of the $NiCo_2O_4$ products obtained by hydrothermal process are shown in **Figure 7** [20].

3. Mechanism and rules of morphology transition engineering

On the basis of chemical and physical explanation we will illustrate the mechanism of morphology transition of pure zinc oxide (ZnO) nanorods to nanoplatelets grafted Mo_8O_{23}-MoO_2 mixed oxide, when react with polyoxoxmetalates (POMs) like phosphomolybdic acid (PMA) under ultra-sonication. In addition, we will discuss a series of selection rules (SRs) which account for the ability of ZnO to experience morphology transition among other binary compounds. The

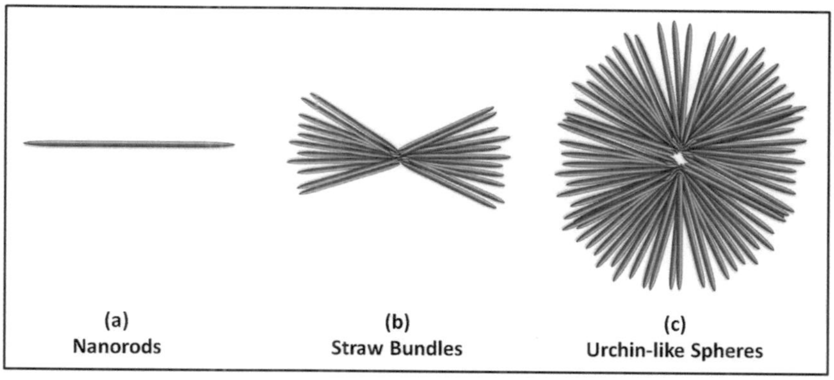

Figure 7.
Shows schematic illustrations of the growth mechanism of the urchin-like $NiCo_2O_4$ nanostructures evolving from nanorods, straw bundles to urchin-like spheres.

explanation of the nano-scale reaction mechanism depends on some theories and models like hard and soft (Lewis) acids and bases (HSAB) theory, Brønsted-Lowry acid/base theory, frontier molecular orbital (FMO) theory, the perturbation theory of reactivity, and site-binding model [21–25]. According to HSAB theory; elements in periodic table can be classified as acids and bases [26]. So we suggest that the cationic metal centers (Zn^{2+}) act as Lewis acid sites [LUMO: Lowest Unoccupied Molecular Orbital] while the anionic oxygen (O^{2-}) centers act as Lewis bases [HOMO: Highest Occupied Molecular Orbital] [27]. We can conclude that the fusion process is due to the interaction between (LUMO) of cations with (HOMO) of ligand (empty-filled interaction) which produce a bonding molecular orbital [28]. When the energy of the interacting orbitals is close the newly formed molecular orbital needs greater stabilizing energy to be formed [29]. Hence, we introduce the first selection rule (SR1) as follow "The appropriate energy difference between LUMO (acid orbital) and HOMO (base orbital) may account for ability of ZnO to experience morphology transition when react with POMs under specific conditions" [7].

Acid/base sites are quite important in determining the catalytic activity of metal oxides [27]. In highly acidic medium (PMA solution (pH ~2)), ZnO surface will act as Brønsted bases and can accept protons according to site-binding model and become positively charged as illustrated in **Figure 8b**. The isoelectric point (IEP) is the point at which the MO carries zero charge [30]. At pH values above the IEP, the MO carries a negative charge, while at pH values below the IEP; the MO carries a positive charge [31]. The IEP of ZnO lies within the pH range (8.0–10.0) [32]. The positively charged ZnO nanorods will be surrounded by the Polyoxomatalate anions $[PO_4\ (MoO_3)_{12}]^{3-}]$ as illustrated in **Figure 8a–c**. Hence, by considering these two factors which are; at pH value lower than the IEP, OH^{2+} species are dominant, and our reaction takes place at extremely acidic medium (pH ~ 2); we introduce the second selection rule (SR2) as follow; "The suitable isoelectric point (IEP) which locates in basic region (8–10) account for the tendency of ZnO to experience morphology transition when react with POMs, as this guarantees the predominant of OH^{2+} species which attract POM anions to metal oxide surface" [7].

Surface polarity helps to explain the morphology transition of ZnO, which refer to "the status of the surface plane whether it contains a stoichiometric ratio of cations (Zn^{2+}) and anions (O^{2-}) or not, which refer to non-polar or polar surface

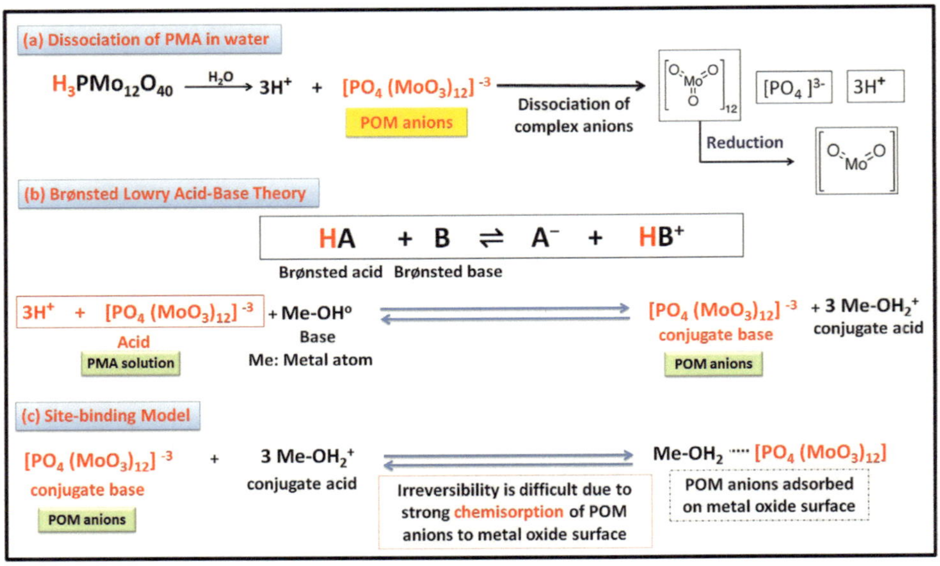

Figure 8.
Shows (a) dissociation equation for phosphomolybdic acid (PMA) in water, (b) the behavior of OH group on MO surface as Brønsted base site (accept protons), and (c) the adsorption of (POM anions) to amphoteric MO surface according to site-binding model. Reproduced with permission [7] 2017, Springer Nature.

respectively" [33]. Physical explanation was suggested on the basis of 'the perturbation theory of reactivity' which state that as a pair of reactants approach one another, their orbitals interact and begin to undergo a "perturbation". Hence, we suggested that when an acid is adsorbed on the oxide surface a perturbation of the acid-base sites [(Zn^{2+}) and (O^{2-}) sites] take place induces the surface relaxation and surface polarity [25]. The polarity may be induced more by chemical etching of surface atoms by PMA. Tasker et al. and other researchers reported that the unstable polar surfaces can attain stability by different methods like metallization (fusion of nanorods), surface reconstruction, adsorption of hydroxyl groups and other charged species, finally further growth occurred due to the presence of other reactants [34–38]. To sum up the mechanism of morphology transition engineering, we conclude three steps for complete morphology transition of ZnO nanorods to hybrid nanoplatelets. First step, involves the chemisorption of POM anions on the surface of ZnO nanorods cause surface relaxation by disrupting the neighboring atoms, which induce the polarity of the non-polar faces (the physical explanation) [7, 9, 11]. According to ZnO Pourbaix diagram, the pH of POM solution (~2) allows the dissolution/chemical etching of ZnO [39]. This chemical etching causes surface relaxation and induces polarity by removing zinc cations or oxygen anions from the outer surfaces (the chemical explanation). These etched species are taking part in self-assembly (fusion) process. Second step involves three main mechanisms; the first one, is the metallization process which involves fusing of two faces with different polarity by bonding Zn-atom to O-atom through their dangling bonds [empty Zn 4 s-orbitals and filled O 2p-orbitals] to form bonding molecular orbital with lower energy and high stability [36]. The second and third ones involve the incorporation of dissolved Zn^{2+} and O^{2-} species within the fused faces which have similar polarity. Schematic illustration for the three possible fusion mechanisms is illustrated in **Figure 9a** [7]. We supposed that PMA acts as a catalyst for binding ZnO nanorods together to nanoplatelets accompanied with deposition of Mo_8O_{23}-MoO_2 mixed oxide on their surfaces. Both of (Mo^{IV}) and (Mo^{VI}) species are oxidized in the air to Mo_8O_{23}-MoO_2 mixed oxide. A brief schematic illustration for proposed morphology transition mechanism is shown in **Figure 9b** [7]. We suggest the formation of intermediate compounds Zn-molybdates (ZM) or/and Zn phosphomolybdate (ZMP) (pigments) during morphology transition process [9, 11].

In addition, we have suggested that, magnetic properties of cations that form the binary compounds may account well for the fluctuation of their behavior when

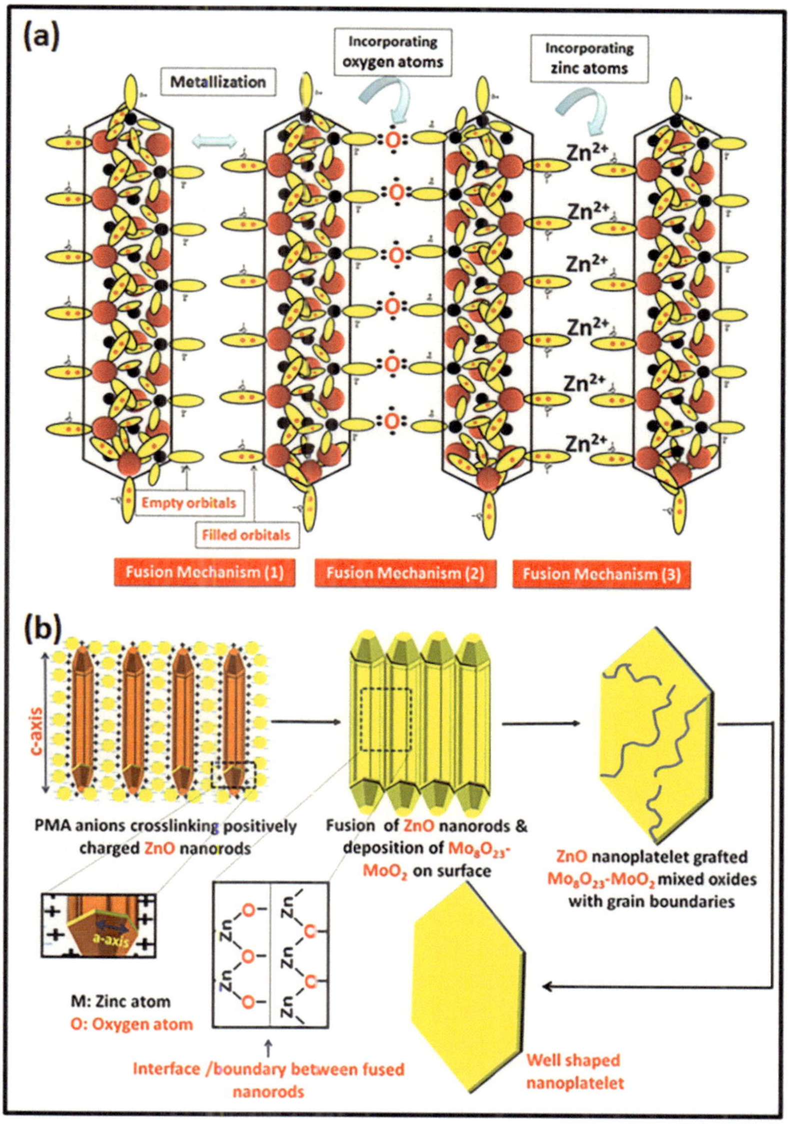

Figure 9.
(a) Schematic illustration for the three possible fusion mechanisms of ZnO nanorods in PMA solution, (b) the proposed mechanism for the morphology transition (self-assembly) of hexagonal ZnO nanorods to ZnO nanoplatelets grafted Mo_8O_{23}-MoO_2 mixed oxide. Reproduced with permission [7] 2017, Springer Nature.

react with POMs [40–42], as by increasing the para-magnestism of cation, the repulsion between the d-orbital of cation and 2p-orbital of anion also increases which does not allow the binding of the latter with 4s-orbital of cations, and restrict the morphology transition process. This may account for the different behavior of metal oxides when react with POMs. We introduce the third selection rule (SR3); as follow "The completely filled 3d-orbital (diamagnetism) of ZnO can account well for its ability to experience morphology transition when react with POMs and recommend other binary compounds of cations like (Cd^{2+}, Hg^{2+}, Cu^{+}, Au^{+}, Ag^{+}) to obey ATMTE" [7]. This third rule still needs more computational studies and experimental evidences to be clear. Amphoteric behavior of ZnO accounts very well for the phenomenon of morphology transition when react with POMs [30, 43]. Amphoterism of ZnO means that, it has the tendency to dissolve in both acidic and basic medium, which guarantees the presence of dissolved cationic and anionic species during all stages of reactions. These species play a critical role in fixing the grain boundaries between the merged nanorods to produce nano-platelets. Hence, we introduce the fourth selection rule (SR4) as follows; "The amphoteric nature of ZnO accounts well for its ability to experience morphology transition when react with POMs at specific conditions, and recommend other amphoteric oxides to obey ATMTE" [7]. On the basis of experimental evidences we have postulated the morphology engineering rules (MERs) that control the morphology (dimension and thickness) of fabricated materials [9]. We introduce the first MER (MER1) as follow, "Under controlled conditions, as the temperature increases, the possibility to engineer higher dimensions (2D,3D) decreases and vice versa" [9], while the second MER (MER2) "Thickness Rule" states that, "Under controlled conditions, ultra-sonication at low temperature and POM concentrations increases the possibility to engineer higher dimensions (2D,3D) with few atoms in thickness" [9]. The detailed mechanism and discussion about the rules are illustrated in [7, 9, 11].

4. Abdelmohsen et al. theories and hypothesis for morphology transition engineering (ATMTE, ATMESC and AHEMNS)

Theory defined as "the explanation of the general causes of certain phenomenon with evidences or facts to support it" [44]. So, on the basis of our phenomenon that discussed in Section 3 and other experimental evidences that will be mentioned later, we can postulate a theory about morphology transition engineering of micro and nanomaterials. Abdelmohsen et al. theory for morphology transition engineering (ATMTE) was proposed on the basis of the phenomenon of morphology transition of ZnO nanorods to nanoplatelets when reacts with POMs under ultra-sonication [7]. This theory (ATMTE) states that "Binary compounds especially amphoteric/diamagnetic pure and doped metal-oxides like (ZnO, Cu_2O) that have appropriate energy difference between their LUMO (acid site/cation) and HOMO (base site/anion), may experience morphology transition to various dimensions (1D, 2D and 3D) when reacts with polyoxometalates under specific conditions, with a possibility to manipulate their surface catalytic properties" [7]. By further study and reading about the morphology transition of materials, another theory was proposed which discuss the ability of most solid materials to experience morphology transition after inducing their surface polarity [11]. This theory is Abdelmohsen et al. theory for morphology engineering of solid compounds (ATMESC), which state that "Most solid compounds may experience morphology transition by inducing their surface polarity (creating dangling bonds); chemically, physically or mechanically. For instance, chemically we can induce polarity of amphoteric compounds which include diamagnetic cations (d^{10}) within their structure (e.g., doped/pure ZnO or Cu_2O and their hydroxides, and composites) with a possibility to manipulate the engineered morphologies, and surface catalytic activity depending on the morphology engineering rules (MERs)" [11]. A hypothetical study for the morphology transition of Cu_2O and other solid materials was proposed as Cu_2O dissolves completely by POMs at high temperature but still need a precipitating-agent to be precipitated to different morphologies. In order to control this reaction we proposed Abdelmohsen et al. hypothesis for engineering of micro and nanostructures (AHEMNS). This hypothesis supposes dissolving any chemical compound by a specific dissolving agent, such as acids, e.g., PMA. Then these dissolved seeds can be precipitated by an alkali like NaOH to synthesize various nanostructures. It is a modification for the aforementioned two theories (ATMTE and ATMESC) and is expected to be applied for nearly all solid compounds by using different dissolving and precipitating-agents [9]. This hypothesis (AHEMNS) states that "Most Solid compounds may be engineered by dissolving starting materials by dissolving agents (e.g., phosphomolybdic acids) which acts also as structural directing-agent, followed by co-precipitating by precipitating agent like (NaOH)" [9]. The produced materials by this method are mostly hybrid materials due to the co-precipitation of both the precursor and the dissolving agent together. In addition, we assumed the fabrication of different morphologies like rods, wires, and needles. Also we look for fabrication of 3D nanostructures when carbon fillers like graphene, and carbon nanotubes, are added during synthesizing. The used dissolving agents is not limited to PMA, as we assume also other types of POMs which widely used for self-assembling of materials and other types of acids to be used as structural-directing agents. The growth of MO over carbonaceous materials is strongly recommended for critical applications like energy storage as this enhance electron diffusion through the whole structure. Two samples of ZnO nanoparticles which are (ZnO (1)) and (ZnO (2)) were prepared by a simple co-precipitation at room temperature and micro-wave assisted technique as reported here. XRD charts and SEM images for them are shown in **Figure 10** [9].

4.1 Fabrication of 1D nanostructures

1D nanostructured materials possess a variety of advantages which make them promising materials in different applications like alkali metal ion batteries due to the high capacity, long-term cycling, and superior rate performance. They also used in electronics, water splitting, photocatalysts, and water treatment [45]. On the basis on our theories, we have fabricated various 1D nanostructures as reported here [7, 9]. For instance, by stirring of ZnO (2) at high temperature (~90°C) in POM solution, a hybrid ZnO nanorods (ZnO grafted MoO_x) was produced which have a diameter in the range of (100–130) nm and length of few micrometer [9]. Few nanosheets are formed but the predominant structure is the nanorods. SEM images of the nanomaterial are shown in **Figure 11** [9]. As noticed by experiments, at room temperature Cu_2O has not experienced any morphology transition, but under ultra-sonication at high temperature (~90°C) hybrid nanowires were produced by addition of NaOH as a precipitating-agent as reported here [9]. XRD charts and SEM images are shown in **Figure 12** [9]. A hybrid (1D) ZnO/MWCNTs nanocomposite was produced by

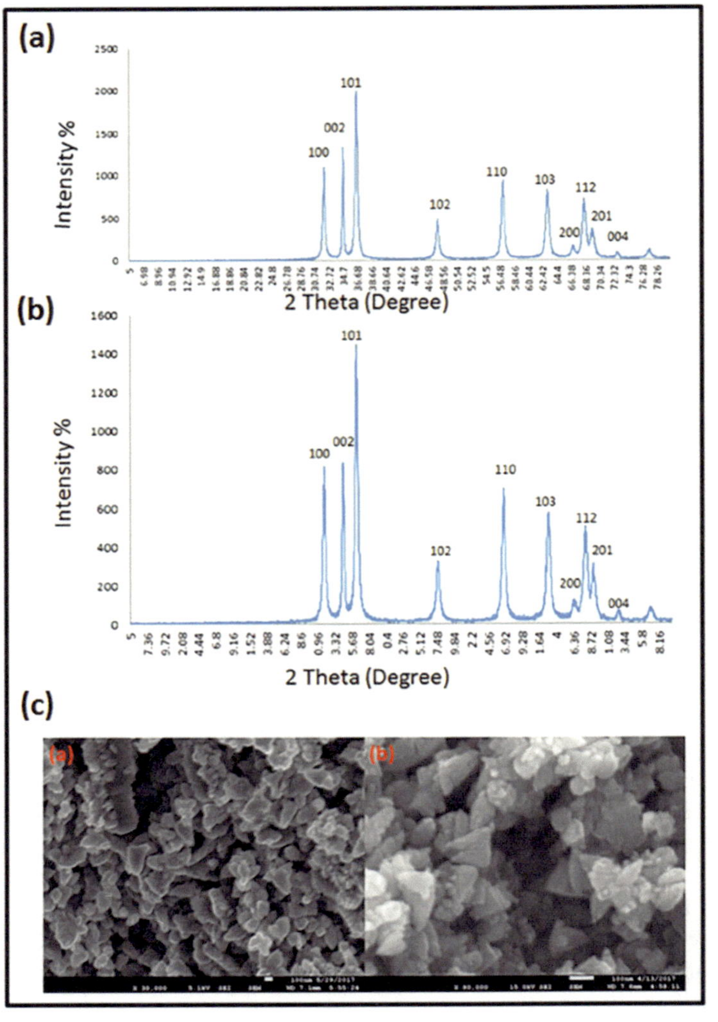

Figure 10.
XRD charts for (a) (ZnO (1)), (b) (ZnO (2)), (c) SEM images for (ZnO (1)) (left) and (ZnO (2)) (right). Reproduced with permission [9] 2019, Springer Nature.

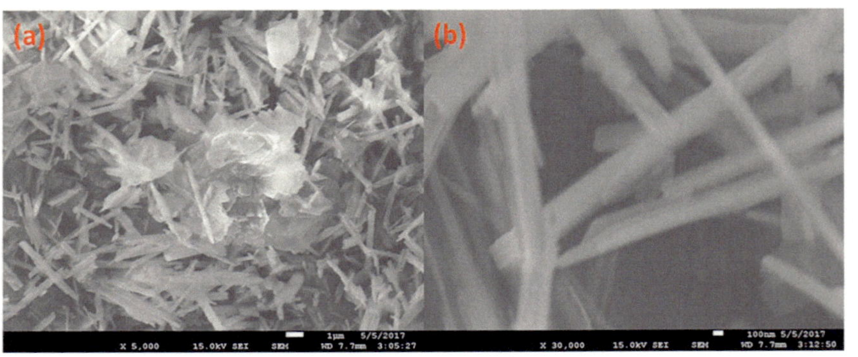

Figure 11.
SEM images of hybrid ZnO nanosheets prepared by stirring of "ZnO (2)" in polyoxometalate (POM) solution at ~90°C. reproduced with permission [9] 2019, Springer Nature.

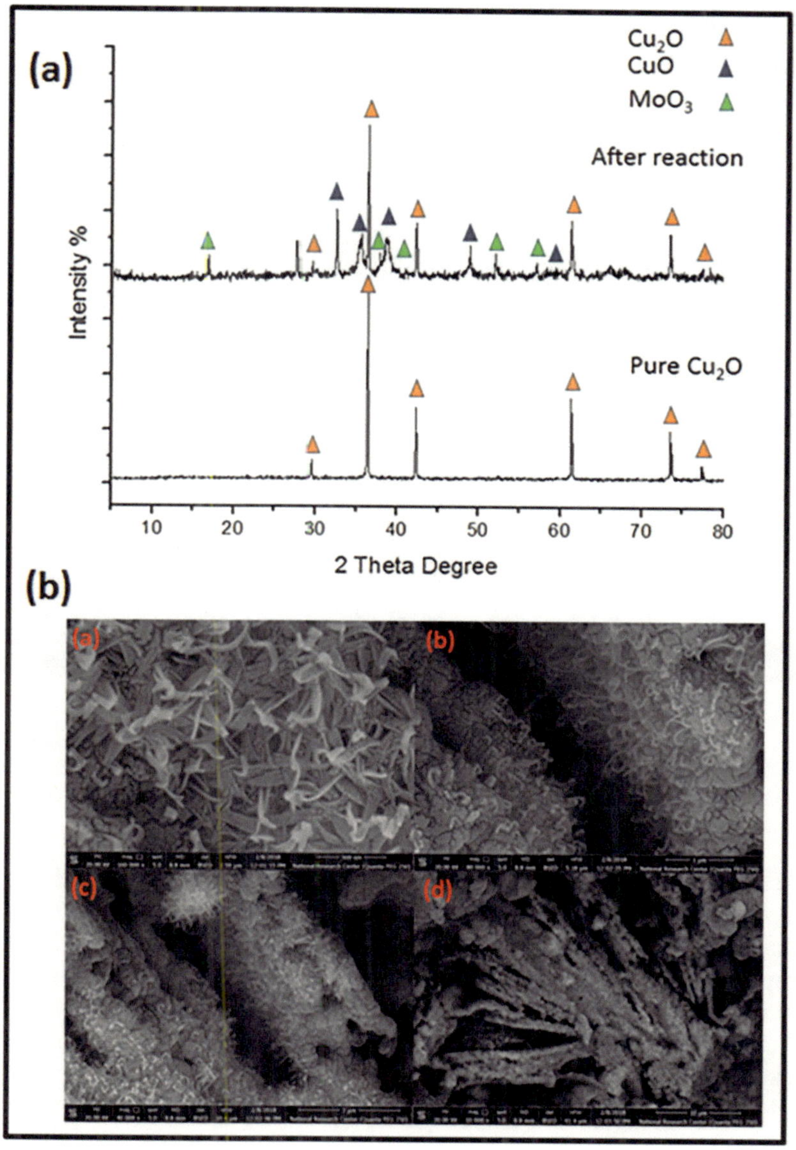

Figure 12.
(a) XRD charts for pure Cu_2O and the formed materials after reaction under ultra-sonication, (b) SEM images of Cu_2O that experienced morphology transition to nanowires under ultra-sonication conditions at high temperature. Reproduced with permission [9] 2019, Springer Nature.

ultra-sonication at room temperature of mixture of ZnO and MWCNTs in POM solution [9]. This will be discussed in details in Section 5.

4.2 Fabrication of 2D nanostructures

2D nanostructured materials possess a variety of advantages which make them promising materials in different applications like energy storage and conversion. They also used in cancer therapy, biomedical applications, piezophototronics, and water treatment [46, 47]. On the basis of our theories, we have fabricated various 2D

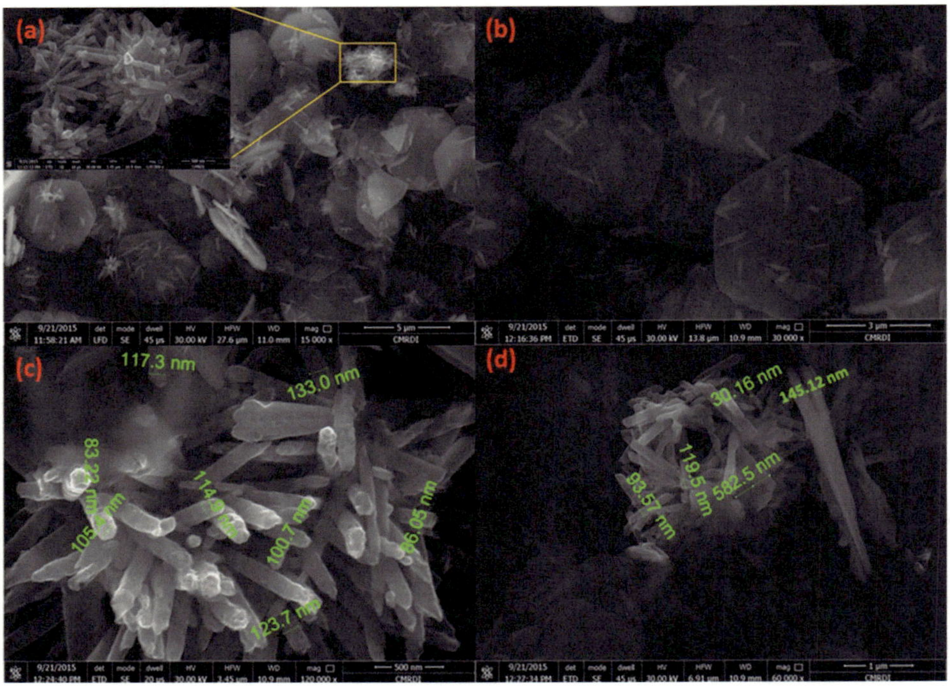

Figure 13.
(a) FESEM images of ZnO nanoflower over nanoplatelets, (b) dispersed ZnO nanorods over nanoplatelets, (c) collection of ZnO nanorods, (d) and ZnO nanocage over nanoplatelets. Reproduced with permission [7] 2017, Springer Nature.

nanostructures as reported here [7, 9]. For instance, we have fabricated this hybrid nanocomposite (ZnO nanoplatelets grafted Mo_8O_{23}-MoO_2 mixed oxide) decorated ZnO nanostructures, by ultra-sonication a mixture of ZnO nanorods and POM at room temperature for nearly 15 minutes. Nanoplatelets with thickness (100–130) nm and micrometer length were produced as shown in **Figure 13** [7]. Under different conditions, thin nanosheets (exfoliated graphite-like structure) were produced with thickness (38–50) nm and micrometer length by also ultra-sonication of ZnO (2) in POM solution at room temperature as shown in **Figure 14** [9]. Moreover, 2D nanosheets were also fabricated by stirring of "ZnO (1)" in polyoxometalate solution at ~90°C as reported here [9].

4.3 Fabrication of 3D nanostructures

3D nanostructured materials possess a variety of advantages which make them promising materials in different applications like electronics, environmental applications, energy storage, and energy conversion [47]. On the basis on our theories, we have fabricated various 3D nanostructures as reported here [9]. For instance, we have fabricated intersected (3D) hybrid ZnO nanosheets/nanoplatelets (ZnO grafted MoO_x) by ultra-sonication of pure ZnO nanoparticles in POM solution as shown in **Figure 15** [9]. In addition, 3D hybrid Cu_2O nanoflowers were fabricated by stirring of Cu_2O in POM solution at high temperature [9]. XRD charts of the starting material (Cu_2O) and the final product is shown in **Figure 16a** [9]. SEM images of the 3D nanoflowers are shown in **Figure 16b** [9].

DOI: http://dx.doi.org/10.5772/intechopen.112135

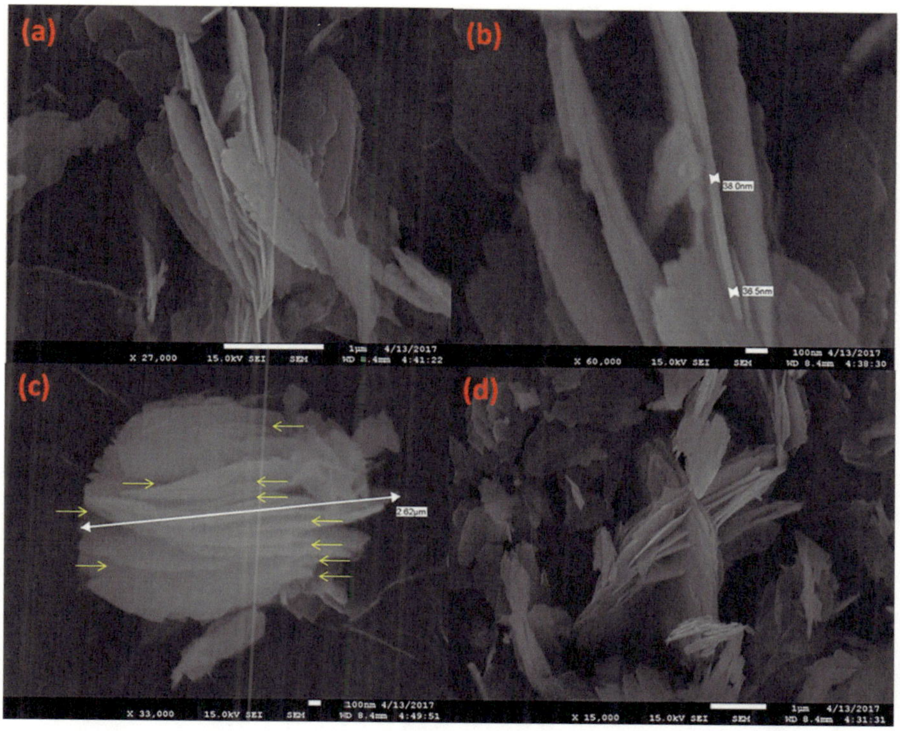

Figure 14.
SEM images of hybrid ZnO multi-layered nanosheets (exfoliated graphite-like structure) prepared by ultra-sonication of "ZnO (2)" at room temperature. Reproduced with permission [9] 2019, Springer Nature.

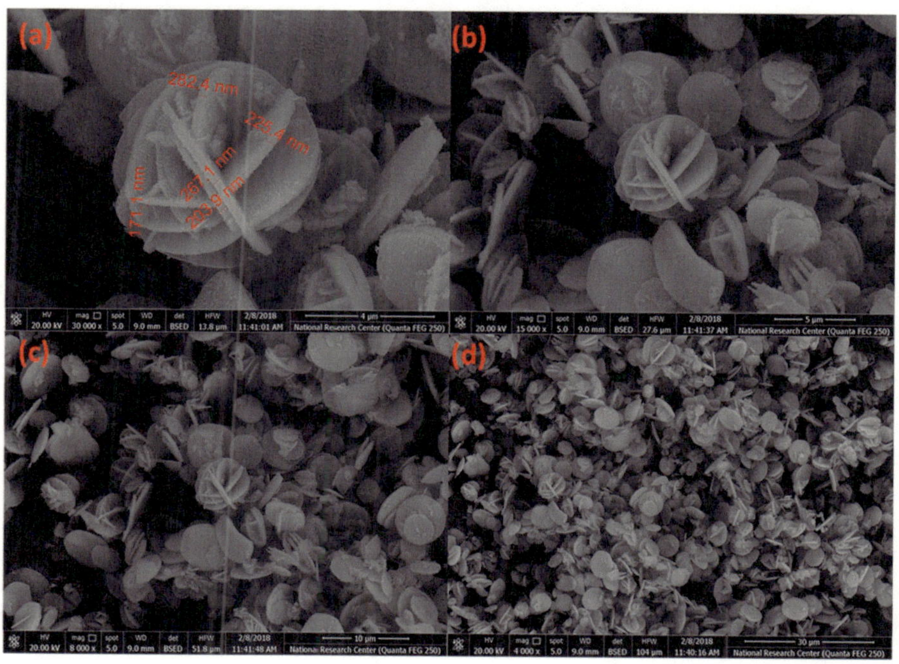

Figure 15.
SEM images of 3D- hybrid ZnO nanostructures prepared by ultra- sonication method. Reproduced with permission [9] 2019, Springer Nature.

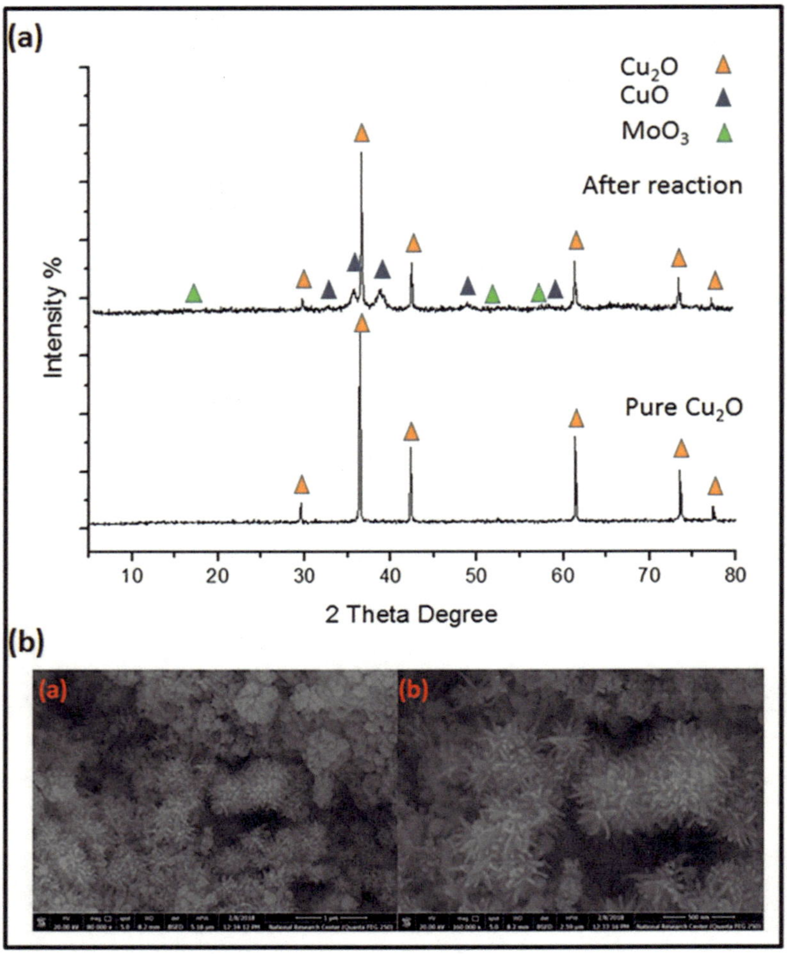

Figure 16.
(a) XRD charts of the starting material (Cu_2O) and the final product, (b) SEM images of the formed 3D nanoflowers. Reproduced with permission [9] 2019, Springer Nature.

5. Influence of carbonaceous materials on morphology transition of ZnO nanoparticles

Carbon-based materials which called the carbon fillers are used for enhancing the mechanical and electrical properties of the nanocomposite. They can be classified in to two types; which are the traditional fillers and the advanced fillers. The traditional fillers are carbon black and graphite. On the other hand, the advanced fillers are expanded graphite, graphene derivatives (graphene oxide, graphene, and porous graphene), carbon nanotubes (multi-walled carbon nanotubes and single-walled carbon nanotubes) and carbon aerogel [48]. These carbon fillers are acting as structural-directing materials which control the structure and morphology of hybrid ZnO during morphology transition process in the presence of POMs which act as structural-directing agent [9]. In our experiment, the effect of the presence of carbon fillers like carbon nanotubes (MWCNTs) was investigated. We have added (0.25 g) of both pure ZnO and MWCNTs in 50 ml deionized water under ultra-sonication for 45 minutes in the presence of PMA at 25°C. Under this reaction conditions we expected the fabrication of nanoplatelets (nanosheets) as usual, but we get a mixture of nanorods and nanosheets. For this unexpected phenomenon we have assumed this mechanism. We assumed that the carbon filler (MWCNTs) which act as structural-directing materials for the polar ZnO nanoparticles can inhibit the dangling bonds on ZnO surfaces that are in contact with them. By this way, they can allow the fusion of ZnO nanoparticles in only one-dimension. This leads to the formation of 1D nanomaterial which is the nanorods. In other words, due to the high strength of MWCNTS, we supposed the ability of them to control the way by which the ZnO nanoparticles will fuse with each other. Growth mechanism of hybrid ZnO nanorods in the presence of multi-walled carbon nanotubes is shown in **Figure 17** [9]. SEM images of hybrid ZnO/multi-walled carbon nanotube nanocomposites are shown in **Figure 18** [9].

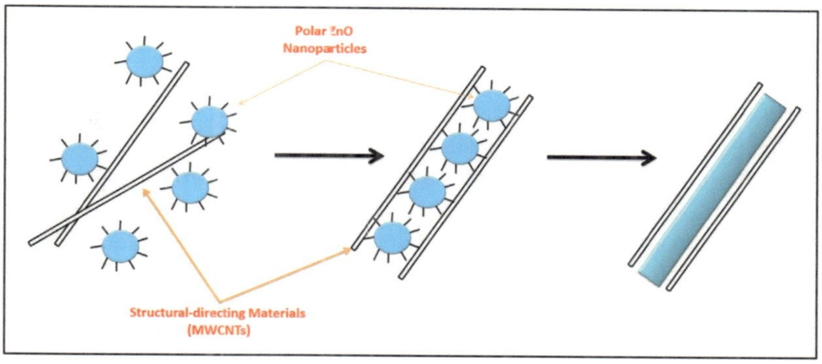

Figure 17.
Growth mechanism of hybrid ZnO nanorods in the presence of MWCNTS.

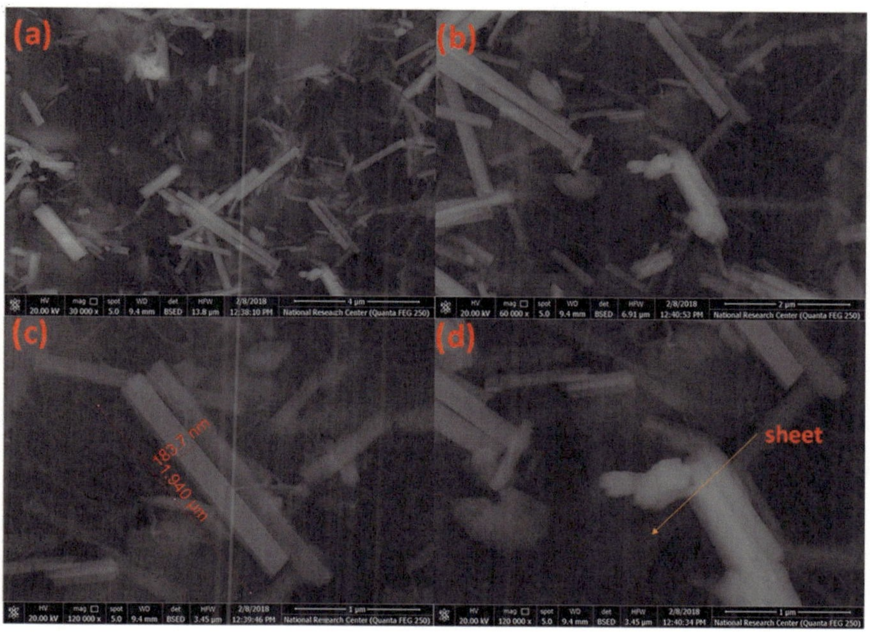

Figure 18.
SEM images of hybrid ZnO/ MWCNTS nanocomposites. Reproduced with permission [9] 2019, Springer Nature.

6. Zinc-oxidene, double layer thin film and expected applications

We suggested the controlling of nanosheets thickness to few nanometers by controlling the reaction conditions like temperature, sonication power and concentration. This will help us to engineer the quantum nanosheets (few/one-atoms thick layer) of hybrid ZnO, which was nominated as hybrid zinc-oxidene [9, 49, 50]. By similar or different methods metal-oxidene can be prepared for a variety of MO. Pure ZnO nanosheets can be produced by chemical etching of MoO_x deposited layers on ZnO nanosheets as shown in **Figure 19a** [9]. Hybrid and pure Zn-oxidene is expected to have extraordinary physico-chemical properties similar to that of graphene and silicone [51]. Fabrication of double layers or hybrid ZnO thin film is conceivable by soaking up of ZnO thin film in POM solution under specific conditions. This will lead to the polarity inducing of top surface, then the deposition of MoO_x layer epitaxially over ZnO substrate as shown in **Figure 19b** [9]. Hybrid and pure 1D, 2D and 3D nanomaterials that are engineered by Abdelmohsen et al. theories (ATMTE, ATMESC, and AHEMNS) are expected to have potential applications in different field if science likes solar cells that strongly depend on thickness [52]. Expected high piezoelectricity of ultra-thin ZnO nanosheets/nanoplatelets will make it an excellent candidate for nanosensors and nanoactuators [53]. ZnO is a promising anode material for alkali metal-ion batteries like lithium-ion battery, sodium-ion batteries and potassium-ion batteries, because it is not expensive, environmentally friendly, and biocompatible [54]. Pure ZnO has a high theoretical capacity (978 mAh g^{-1}) and can be used in its pure form as anode or doped with other oxides like transition metal oxides such as CoO (715 mAh g^{-1}), NiO (718 mAh g^{-1}) and CuO (674 mAh g^{-1}). Pure and doped ZnO nanostructures are also used as an electrode in supercapacitors [55].

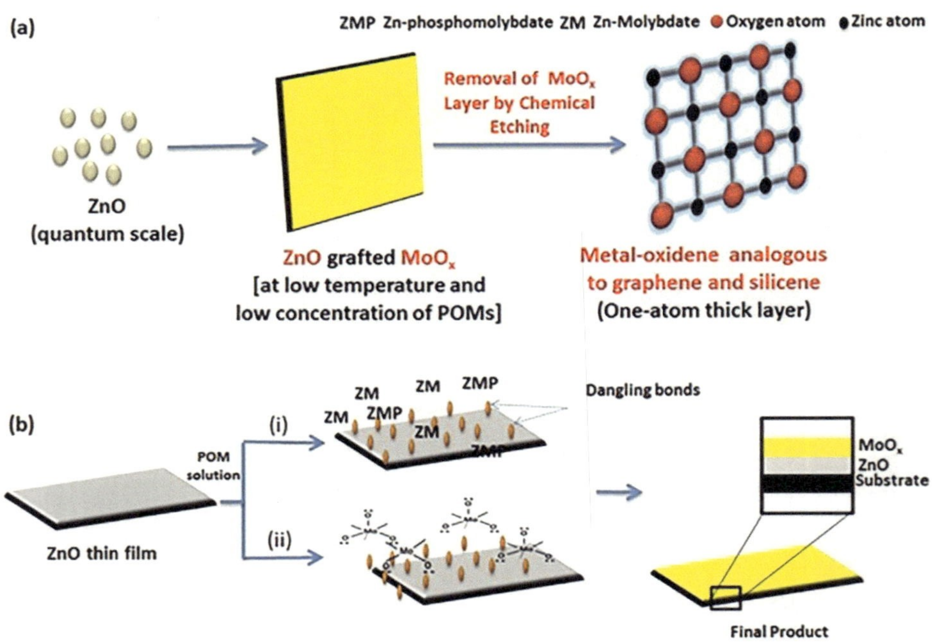

Figure 19.
(a) Schematic illustration for developing thin sheet of ZnO (Zn-oxidene) by chemical etching of MoO_x, and (b) developing thin film of MoO_x/ZnO top-down layers. Reproduced with permission [9] 2019, Springer Nature.

DOI: http://dx.doi.org/10.5772/intechopen.112135

Other applications for pure ZnO are rubber industry, ceramic industry, medicine, food additive, pigment, UV absorber, coatings, corrosion prevention in nuclear reactors, methane reforming, and electronics [56].

7. Conclusion

These theories are expected to be the basis for reliable engineering of hybrid nanostructures with different dimensions and their functionality. This approach is expected to be the basis of a competitive fabrication approach to 1D, 2D and 3D hybrid nanostructures. These hybrid nanomaterials have potential use for energy, catalysis, biomedical and other applications. Additional experimental evidences are required to support these theories. Further work is required to manipulate the thickness of nanoplatelets/nanosheets, study catalytic properties of hybrid materials, and study morphology transition in the presence of carbonaceous materials (Graphene, CNTs, and CNPs) and their applications to alkali metal-ion batteries. In addition, more computational studies and experimental evidences are required to support our suggested mechanisms, theories, and the rules. Finally, we have intention to synthesis other ployoxometalates (POMs) and check their ability to engineer morphology transition of zinc oxides and other metal oxides or solid compounds to hybrid nanostructures.

Acknowledgements

We acknowledge the European master program "Functional Advanced materials and Engineering with Artificial Intelligence for Sustainability (FAME$^{(AIS)}$ Master - Erasmus Mundus)" for funding Mr. Ahmed Hashem Abdelmohsen to do his master studies in Augsburg University, Grenoble Alpes University and Catholic University of Louvain in which he postulated these theories, hypothesis and rules about morphology transition engineering of micro and nanomaterials.

Conflict of interest

The authors declare no conflict of interest.

Author details

Ahmed Hashem Abdelmohsen[1,2,3*†], Sherif A. El-Khodary[1,4†] and Nahla Ismail[5]

1 Institute for Energy Research, School of Material Science and Engineering, Jiangsu University, Zhenjiang, Jiangsu, China

2 The Central Laboratory, Faculty of Postgraduate Studies for Advanced Science (PSAS), Beni-Suef University, Beni-Suef, Egypt

3 Institute of Physics, Augsburg University, Augsburg, Germany

4 Building Physics and Environment Institute, Housing and Building National Research Center (HBRC), Cairo, Egypt

5 Physical Chemistry Department, Centre of Excellence for Advanced Sciences, Renewable Energy Group, National Research Centre, Giza, Egypt

*Address all correspondence to: ahmed.h.abdelmohsen@gmail.com

† These authors contributed equally to this work.

References

[1] Abu-Lebdeh Y. Introduction. In: Abu-Lebdeh Y, Davidson I, editors. Nanotechnology for Lithium-Ion Batteries. 1st ed. Canada: Springer; 2013. pp. 1-12. DOI: 10.1007/978-1-4614-4605-7.ch1

[2] Asha A, Narain R. Nanomaterials properties. In: Narain R, editor. Polymer Science and Nanotechnology: Fundamentals and Applications. Canada: Elsevier; 2020. pp. 343-359. DOI: 10.1016/B978-0-12-816806-6.00015-7.ch15

[3] Carnovale C, Bryant G, Shukla R, Bansal V. Size, shape and surface chemistry of nano-gold dictate its cellular interactions, uptake and toxicity. Progress in Materials Science. 2016;**83**:152-190. DOI: 10.1016/j.pmatsci.2016.04.003

[4] Pereira L. Investigating mechanical properties and thermal conductivity of 2D carbon-based materials by computational experiments. Computational Materials Science. 2021;**196**:110493. DOI: 10.1016/j.commatsci.2021.110493

[5] Sharifian A, Baghani M, Wu J, Odegard G, Baniassadi M. Insight into geometry-controlled mechanical properties of spiral carbon-based nanostructures. Journal of Physical Chemistry C. 2019;**123**:3226-3238. DOI: 10.1021/acs.jpcc.8b12269

[6] Vanmaekelbergh D, Houtepen A, Kelly J. Electrochemical gating: A method to tune and monitor the (opto) electronic properties of functional materials. Electrochimica Acta. 2007;**53**:1140-1149. DOI: 10.1016/j.electacta.2007.02.045

[7] Abdelmohsen A, Rouby W, Ismail N, Farghali A. Morphology transition engineering of ZnO nanorods to nanoplatelets grafted Mo_8O_{23}-MoO_2 by polyoxometalates: Mechanism and possible applicability to other oxides. Scientific Reports. 2017;**7**:1-26. DOI: 10.1038/s41598-017-05750-x

[8] Abid N, Khan A, Shujait S, Chaudhary K, Ikram M, Imran M, et al. Synthesis of nanomaterials using various top-down and bottom-up approaches, influencing factors, advantages, and disadvantages: A review. Advances in Colloid and Interface Science. 2022;**300**:102597. DOI: 10.1016/j.cis.2021.102597

[9] Abdelmohsen A, Ismail N. Morphology transition of ZnO and Cu_2O nanoparticles to 1D,2D, and 3D nanostructures: Hypothesis for engineering of micro and nanostructures (HEMNS). JSST. 2020;**94**:213-228. DOI: 10.1007/s10971-019-05114-z

[10] Lannoo M. The role of dangling bonds in the properties of surfaces and interfaces of semiconductors. Review Physics Applied. 1990;**25**:887-894. DOI: 10.1051/rphysap:01990002509088700

[11] Abdelmohsen AH. Theory for morphology engineering of solid compounds (ATMESC): Facts, predictions and experimental evidences. Journal of Material Science Engineering. 2017;**6**:377. DOI: 10.4172/2169-0022.1000377

[12] Basset D, Matteazzi P, Miani F. Designing a high energy ball-mill for synthesis of nanophase materials in large quantities. Materials Science Engineering: A. 1993;**168**:149-152. DOI: 10.1016/0921-5093(93)90718-T

[13] Hashizume T, Heike S, Lutwyche MI, Watanabe S, Nakajima K. Interaction of Ga adsorbates with dangling bonds on the hydrogen terminated Si (100) surface. Japanese Journal of Applied Physics. 1996;**35**:L1085. DOI: 10.1143/JJAP.35.L1085

[14] Wang L, Liu G, Zou L, Xue D. Phase evolution from rod-like ZnO to plate-like zinc hydroxysulfate during electrochemical deposition. Journal of Alloys and Compounds. 2010;**493**:471-475. DOI: 10.1016/j.jallcom.2009.12.129

[15] Yerra S, Supriya S, Das S. Reversible morphological transition between nano-rods to micro-flowers through micro-hexagonal crystals in a sonochemical synthesis based on polyoxovanadate compound. Inorganic Chemistry Communications. 2013;**35**:54-57. DOI: 10.1016/j.inoche.2013.04.029

[16] Xiong S, Yuan C, Zhang X, Xi B, Qian Y. Controllable synthesis of mesoporous Co_3O_4 nanostructures with tunable morphology for application in Supercapacitors. Chemistry - A European Journal. 2009;**15**:5320-5326. DOI: 10.1002/chem.200802671

[17] Yan L, Yu R, Chen J, Xing X. Template-free hydrothermal synthesis of CeO_2 Nano-octahedrons and nanorods: Investigation of the morphology evolution. Crystal Growth & Design. 2008;**8**:1474-1477. DOI: 10.1021/cg800117v

[18] Gao Y, Nagai M. Morphology evolution of ZnO thin films from aqueous solutions and their application to solar cells. Langmuir. 2006;**22**:3936-3940. DOI: 10.1021/la053042f

[19] Sweegers C, Coninck H, Meekes H, van Enckevort W, Hiralal I, Rijkeboer A. Morphology, evolution and other characteristics of gibbsite crystals grown from pure and impure aqueous sodium aluminate solutions. Journal of Crystal Growth. 2001;**233**:567-582. DOI: 10.1016/S0022-0248(01)01615-3

[20] Wang Q, Liu B, Wang X, Ran S, Wang L, Chen D, et al. Morphology evolution of urchin-like $NiCo_2O_4$ nanostructures and their applications as psuedocapacitors and photoelectrochemical cells. Journal of Materials Chemistry. 2012;**22**:21647. DOI: 10.1039/C2JM34705A

[21] Pearson R. Recent advances in the concept of hard and soft acids and bases. Journal of Chemical Education. 1987;**64**:561. DOI: 10.1021/ed064p561

[22] Li Y, Evans J. The Fukui function – A key concept linking frontier molecular-orbital theory and the hard-soft-acid-base principle. JACS. 1995;**117**:7756-7759. DOI: 10.1021/ja00134a021

[23] Yatess E, Levine S, Healy T. Site-binding model of the electrical double layer at the oxide/water interface. Journal of the Chemical Society, Faraday Transactions. 1974;**1**:1807-1818. DOI: 10.1039/F19747001807

[24] Wiśniewska M, Chibowski S, Urban T. Comparison of adsorption affinity of ionic polyacrylamide for the surfaces of selected metal oxides. Adsorption Science Technology. 2017;**35**:582-591. DOI: 10.1177/0263617417702385

[25] Zecchina A, Lamberti C, Bordiga S. Surface acidity and basicity: General concepts. Catalysis Today. 1998;**41**:169-177. DOI: 10.1016/S0920-5861(98)00047-9

[26] Jolly W. Modern Inorganic Chemistry. USA: Mcgraw Hill International Edition; 2017

[27] Abee M. Interaction of Acid/Base Probe Molecules with Specific Features on Well-Defined Metal Oxide Single-Crystal Surfaces. Blacksburg, Virginia; 2001

[28] Glazneva T, Kotsarenko N, Paukshtis E. Surface acidity and basicity of oxide catalysts: From aqueous suspensions to in situ measurements. Kinetics and Catalysis. 2008;**49**:855-867. DOI: 10.1134/S0023158408060104

[29] Jean Y. Molecular Orbitals of Transition Metal Complexes. UK: Oxford University Press; 2005

[30] Zhu B, Xia P, Ho W, Yu J. Isoelectric point and adsorption activity of porous gC_3N_4. Applied Surface Science. 2015;**344**:188-195. DOI: 10.1016/j.apsusc.2015.03.086

[31] Brunello J. Preparation of catalysts by metallic complex adsorption on mineral oxides. Pure Applied Chemistry. 1978;**50**:1211-1229. DOI: 10.1351/pac197850091211

[32] Parks G. The isoelectric points of solid oxides, solid hydroxides, and aqueous hydroxo complex systems. Chemical Reviews. 1965;**65**:177-198. DOI: 10.1021/cr60234a002

[33] Goniakowski J, Finocchi F, Noguera C. Polarity of oxide surfaces and nanostructures. Reports on Progress in Physics. 2008;**71**:016501. DOI: 10.1088/0034-4885/71/1/016501

[34] Tasker P. The stability of ionic crystal surfaces. Journal of Physics C: Solid State Physics. 1979;**12**:4977. DOI: 10.1088/0022-3719/12/22/036

[35] Meyer B. First-principles study of the polar O-terminated ZnO surface in thermodynamic equilibrium with oxygen and hydrogen. Physical Review B. 2004;**69**:045416. DOI: 10.1103/PhysRevB.69.045416

[36] Wander A, Schedin F, Steadman P, Norris A, et al. Stability of polar oxide surfaces. Physical Review Letters. 2001;**86**:3811. DOI: 10.1103/PhysRevLett.86.3811

[37] Chang S, Mark P. The crystallography of the principal non-polar (112-0) and (101-0) surfaces of zinc oxide. Surface Science. 1974;**45**:72. DOI: 10.1016/0039-6028(74)90203-9

[38] Dulub O, Diebold U, Kresse G. Novel stabilization mechanism on polar surfaces: ZnO (0001)-Zn. Physical Review Letters. 2003;**90**:016102-1. DOI: 10.1103/PhysRevLett.90.016102

[39] Price P, Walker R. Chemical etching of charged-particle tracks in solids. Journal of Applied Physics. 1962;**33**:3407. DOI: 10.1063/1.1702421

[40] Fang Z, Terakura K. Structural distortion and magnetism in transition metal oxides: Crucial roles of orbital degrees of freedom. Journal of Physics: Condensed Matter. 2002;**14**:3001-3014. DOI: 10.1088/0953-8984/14/11/312

[41] Tokura Y, Nagaosa N. Orbital physics in transition-metal oxides. Science. 2000;**288**:462-468. DOI: 10.1126/science.288.5465.462

[42] Kuehn K. A Student's Guide through the Great Physics Texts Volume III: Electricity, Magnetism and Light. USA; 2016

[43] Laudise R, Ballman A. Hydrothermal synthesis of zinc oxide and zinc sulfide. The Journal of Physical Chemistry. 1960;**6**:688. DOI: 10.1021/j100834a511

[44] Brown T, Eugene H, Bursten B, Murphy C, Woodward P, Stoltzfus M.

Chemistry the Central Science. 13th ed. USA: Theodore Lawrence; 2015. pp. 2-39

[45] Chen C, Fan Y, Gu J, Wu L, Passerini S, Mai L. One-dimensional nanomaterials for energy storage. Journal of Physics D: Applied Physics. 2018;**51**:113002. DOI: 10.1088/1361-6463/aaa98d

[46] Zhang H, Cheng H, Ye P. 2D nanomaterials: Beyond graphene and transition metal dichalcogenides. Chemical Society Reviews. 2018;**47**:6009-6012. DOI: 10.1039/C8CS90084A

[47] Wan K, Li Y, Wang Y, Wei G. Recent advance in the fabrication of 2D and 3D metal carbides-based nanomaterials for energy and environmental applications. Nanomaterials. 2021;**11**:246. DOI: 10.3390/nano11010246

[48] Zhou Z, Wang S, Zhang Y, Zhang Y. Effect of different carbon fillers on the properties of PP composites: Comparison of carbon black with multiwalled carbon nanotubes. Journal of Applied Polymer Science. 2006;**102**:4823-4830. DOI: 10.1002/app.24722

[49] Abdelmohsen A, Bouchet R, Vlad A. Morphology transition engineering of 1D, 2D, and 3D hybrid nanostructures. In: 9th EMMC FAME Research Workshop. Belgium: Liège University, Research Gate; 2017

[50] Abdelmohsen A, Vlad A. A Facile Synthesis of Exfoliated Graphite like Hybrid Zinc Oxide at Room Temperature: First Step towards Metal-oxidene. Belgium: UCLouvain (IMCN), Research Gate; 2017

[51] Hartman T, Sofer Z. Beyond graphene: Chemistry of group 14 graphene analogues: Silicene, Germanene, and Stanene. ACS Nano. 2019;**13**:8566-8576. DOI: 10.1021/acsnano.9b04466

[52] Suliman A, Tang Y, Xu L. Preparation of ZnO nanoparticles and nanosheets and their application to dye-sensitized solar cells. Solid Energy Material Solids. 2007;**9**:1658-1662. DOI: 10.1016/j.solmat.2007.05.014

[53] Wang J. Biomolecule-functionalized nanowires: From nanosensors to Nanocarriers. ChemPhysChem. 2009;**10**:1748-1755. DOI: 10.1002/cphc.200900377

[54] Guo R, Yue W, An Y, Ren Y, Yan X. Graphene-encapsulated porous carbon-ZnO composites as high-performance anode materials for Li-ion batteries. Electrochimica Acta. 2014;**135**:161-167. DOI: 10.1016/j.electacta.2014.04.160

[55] Zhang Y, Li H, Pan L, Lu T, Sun Z. Capacitive behavior of graphene-ZnO composite film for supercapacitors. Journal of Electroanalytical Chemistry. 2009;**634**:68-71. DOI: 10.1016/j.jelechem.2009.07.010

[56] Özgür Ü, Hofstetter D, Morkoç H. ZnO devices and applications: A review of current status and future prospects. IEEE. 2010;**98**:1255-1268. DOI: 10.1109/JPROC.2010.2044550

Section 2

Type of Nanofabrication

Chapter 3

Fabrication of Mesoporous Silica Nanoparticles and Its Applications in Drug Delivery

Vishal Pande, Sachin Kothawade, Sharmila Kuskar, Sandesh Bole and Dinesh Chakole

Abstract

Mesoporous Silica Nanoparticles (MSNs) are nano-sized particles with a porous structure that offers unique advantages for drug delivery systems. The chapter begins with an introduction to MSNs, providing a definition of these nanoparticles along with a brief historical overview. The distinctive properties of MSNs, such as high surface area, tunable pore size, and excellent biocompatibility, are discussed, highlighting their potential in drug delivery applications. The synthesis methods for MSNs are presented, including template-assisted synthesis, sol-gel method, co-condensation method, and other approaches. The chapter also covers the characterization techniques used for evaluating MSNs, including morphological, structural, and chemical characterization, which are crucial for assessing their quality and functionality. The surface modification of MSNs is explored, focusing on the functionalization of surface groups, attachment of targeting ligands, and surface charge modification to enhance their interactions with specific cells or tissues. The chapter then delves into the diverse applications of MSNs, with a particular focus on drug delivery. The use of MSNs in cancer theranostics, drug delivery, imaging, biosensing, and catalysis is discussed, emphasizing their potential to revolutionize these areas. Furthermore, the toxicity and biocompatibility of MSNs are addressed, covering both in vitro and in vivo studies that evaluate their safety and efficacy.

Keywords: mesoporous silica nanoparticles, surface modification, cancer theranostics, template-assisted synthesis, Co-condensation, biosensing

1. Introduction

1.1 Definition of mesoporous silica nanoparticles (MSNs)

Mesoporous silica nanoparticles (MSNs) are nano-sized particles composed of silica (SiO_2) with a unique porous structure. They possess a regular arrangement of mesopores, which are cylindrical channels or voids within the particle structure [1]. These mesopores typically range in size from 2 to 50 nm, offering a large surface area for drug loading and delivery [2].

The term "mesoporous" refers to the intermediate pore size between micropores (less than 2 nm) and macropores (greater than 50 nm). This mesoporous structure allows for the efficient encapsulation, storage, and controlled release of therapeutic agents, making MSNs highly desirable for drug delivery applications [3].

MSNs can be fabricated with precise control over their size, shape, and pore characteristics, enabling tailored drug delivery systems. The synthesis of MSNs involves the use of various templating agents or surfactants that serve as a template around which the silica precursor is deposited, followed by the removal of the template to create the desired pore structure [4].

The unique properties of MSNs make them versatile platforms for drug delivery. Their large surface area facilitates high drug-loading capacity, while the tunable pore size allows for the selective encapsulation of different types of drugs, including small molecules, proteins, nucleic acids, and even imaging agents. Additionally, the surface of MSNs can be functionalized with targeting ligands, making them capable of targeted drug delivery to specific cells or tissues [5].

1.2 Brief history of MSNs

The development and exploration of mesoporous silica nanoparticles (MSNs) as versatile materials for drug delivery applications have evolved over several decades. The following provides a detailed overview of the historical milestones and key contributions in the field of MSNs [6, 7].

1.2.1 Discovery of mesoporous materials

The study of mesoporous materials traces its origins to the late 1960s and early 1970s when researchers discovered the existence of ordered porous structures in various materials. Initial investigations focused on mesoporous materials like MCM-41 and SBA-15, which laid the foundation for the development of MSNs [8].

1.2.2 Introduction of MSNs

The concept of mesoporous silica nanoparticles (MSNs) emerged in the early 1990s when researchers introduced a template-assisted synthesis method to create highly ordered porous structures within silica nanoparticles. This breakthrough led to the realization of MSNs as promising candidates for drug delivery systems due to their unique properties [9].

1.2.3 Pioneering synthesis methods

In the late 1990s and early 2000s, researchers developed and refined various synthesis methods for MSNs. The sol-gel method and co-condensation method gained significant attention as effective approaches to fabricate MSNs with controlled pore size and surface properties. These methods allowed for the synthesis of MSNs with tailored characteristics suitable for drug delivery applications [10].

1.2.4 Advancements in surface modification

During the 2000s, considerable progress was made in surface modification techniques for MSNs. Researchers explored different strategies to functionalize the surface of MSNs with organic groups, polymers, or targeting ligands. These modifications enabled the introduction of specific functionalities, such as controlled drug release, enhanced stability, and targeted drug delivery [11].

1.3 Unique properties of MSNs

Mesoporous silica nanoparticles (MSNs) possess several distinct properties that make them highly desirable for drug delivery applications. Understanding these unique characteristics is crucial for harnessing the full potential of MSNs in the field of drug delivery [12–15]. The following provides a detailed exploration of the key properties of MSNs:

1.3.1 High surface area

MSNs have an exceptionally high surface area due to their porous structure. The presence of mesopores enables a large surface area-to-volume ratio, providing ample space for drug loading and adsorption. The high surface area allows for efficient interaction with drugs, leading to improved encapsulation efficiency and enhanced drug loading capacity.

1.3.2 Tunable pore size and volume

MSNs offer the advantage of tunable pore size and volume, enabling customization based on specific drug delivery requirements. The pore size can be precisely controlled during the synthesis process, allowing for the selective encapsulation of different types of drugs, including small molecules, macromolecules, and even imaging agents. This tunability facilitates the optimization of drug release kinetics and enhances therapeutic efficacy.

1.3.3 Controlled drug release

MSNs exhibit controlled drug release behavior, which is critical for achieving sustained and targeted drug delivery. The porous structure of MSNs provides a reservoir-like effect, allowing for controlled release of encapsulated drugs over an extended period. The release kinetics can be further modulated through surface modifications, such as the introduction of stimuli-responsive systems or functional groups that respond to specific environmental cues.

1.3.4 Excellent biocompatibility

MSNs demonstrate excellent biocompatibility, ensuring their compatibility with biological systems and minimizing adverse effects. The silica material used in MSNs is generally considered biologically inert, reducing the likelihood of cytotoxicity and immunogenicity. Furthermore, MSNs can be surface-modified with biocompatible polymers or targeting ligands to enhance their biocompatibility and reduce potential toxicity concerns.

1.3.5 Surface functionalization

The surface of MSNs can be easily functionalized, allowing for the introduction of specific functionalities to facilitate targeted drug delivery. Functional groups, polymers, or targeting ligands can be attached to the surface of MSNs, enabling selective interactions with specific cells, tissues, or biological molecules. Surface functionalization also enables the incorporation of imaging agents for real-time monitoring or diagnostic purposes.

1.3.6 Stability and ease of fabrication

MSNs exhibit good stability, making them suitable for long-term storage and transportation. The synthesis methods for MSNs are well-established and relatively straightforward, offering reproducibility and scalability. This ease of fabrication facilitates their translation from the laboratory to industrial-scale production for practical applications.

MSNs hold significant potential for the development of advanced drug delivery systems. Their high surface area, tunable pore size, controlled drug release behavior, excellent biocompatibility, surface functionalization capabilities, and stability make them versatile platforms for efficient and targeted drug delivery, ultimately improving therapeutic outcomes [16–20].

2. Synthesis of MSNs

The synthesis of mesoporous silica nanoparticles (MSNs) involves various methods that allow for the controlled fabrication of their unique porous structure [21–23]. The following describes the key synthesis methods commonly employed for the production of MSNs:

2.1 Template-assisted synthesis

Template-assisted synthesis is a widely employed method for fabricating mesoporous silica nanoparticles (MSNs) with precise control over their porous structure [24–26]. This approach involves the use of a sacrificial template or surfactant, around which the silica precursor is deposited and subsequently removed, leaving behind the desired porous architecture within the nanoparticles. The template provides a framework that determines the size, shape, and arrangement of the mesopores, allowing for tailored drug loading and release properties. The following steps outline the template-assisted synthesis process:

2.1.1 Template selection

The choice of template plays a crucial role in determining the characteristics of the resulting MSNs. Commonly used templates include organic surfactants, such as cetyltrimethylammonium bromide (CTAB) or Pluronic block copolymers, and inorganic templates like colloidal silica or polymer micelles. The template's structure and size dictate the dimensions and arrangement of the mesopores within the silica network.

2.1.2 Silica precursor deposition

The silica precursor, often tetraethyl orthosilicate (TEOS) or other alkoxysilanes, is added to a solution containing the template. Under appropriate conditions, hydrolysis and condensation reactions occur, leading to the formation of silica species.

2.1.3 Template removal

After the deposition of the silica precursor, the template is selectively removed to generate the porous structure within the nanoparticles. The template removal can be achieved through either calcination or extraction processes. Calcination involves heating the MSNs to high temperatures, typically above 400°C, to burn off the organic template. On the other hand, extraction involves dissolving the template using solvents that are selective for the template material. This step leaves behind an interconnected network of mesopores within the MSNs.

2.1.4 Post-synthesis treatment

To further refine the MSNs' properties, post-synthesis treatments can be performed. These treatments may include washing with solvents to remove residual impurities, surface functionalization to introduce specific functionalities or targeting ligands, or modification of the pore surface to alter the release kinetics or enhance biocompatibility.

Template-assisted synthesis offers several advantages for MSN fabrication. It allows precise control over the pore size, distribution, and connectivity, enabling the customization of drug loading and release profiles. The use of different templates and adjustments in synthesis conditions enable the creation of MSNs with specific properties tailored for various drug delivery applications. Additionally, template-assisted synthesis is a relatively straightforward and scalable method, making it suitable for large-scale production of MSNs.

By utilizing template-assisted synthesis, researchers can create MSNs with a well-defined mesoporous structure, offering improved drug encapsulation capacity, controlled release behavior, and enhanced therapeutic efficacy. This synthesis approach contributes to the development of advanced drug delivery systems with precise control over drug release kinetics and targeted delivery to specific tissues or cells (**Figure 1**).

2.2 Sol-gel method

The sol-gel method is a widely used approach for the synthesis of mesoporous silica nanoparticles (MSNs) [27–31]. This method involves the hydrolysis and condensation of silica precursors in a solution, leading to the formation of silica nanoparticles with a porous structure. The sol-gel process allows for the precise control of reaction parameters to tailor the size, shape, and pore characteristics of the resulting MSNs. The following steps outline the sol-gel method for MSN synthesis:

2.2.1 Selection of silica precursors

The sol-gel method typically utilizes alkoxysilane precursors, such as tetraethyl orthosilicate (TEOS) or tetramethyl orthosilicate (TMOS). These precursors undergo hydrolysis and subsequent condensation reactions to form the silica network. The

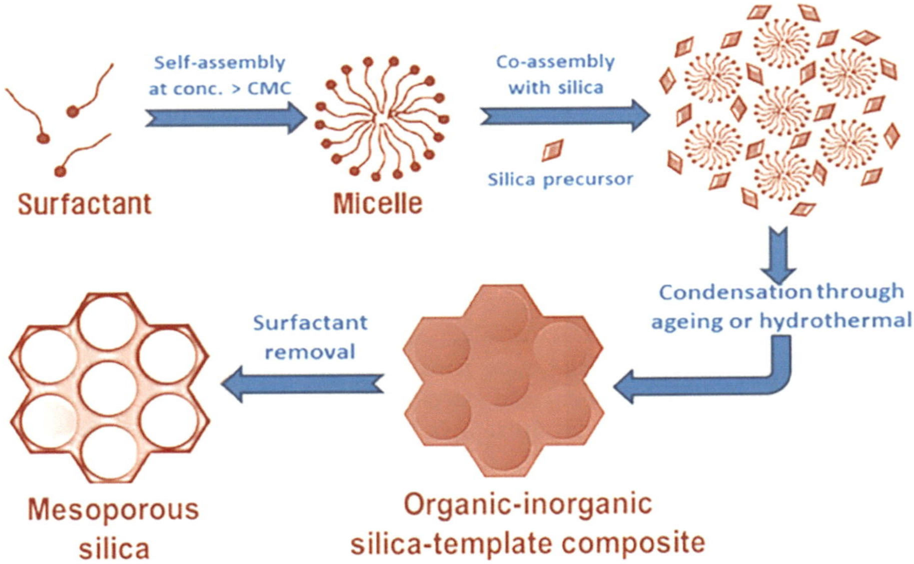

Figure 1.
Template assisted synthesis of mesoporous silica nanoparticles.

choice of precursor depends on factors such as reactivity, availability, and desired properties of the MSNs.

2.2.2 Hydrolysis

The silica precursor is hydrolyzed by adding a controlled amount of water or a hydrolyzing agent, such as an acid or base, to initiate the hydrolysis reaction. This step involves the breaking of the alkoxide groups in the precursor, resulting in the formation of silanol groups (Si-OH) on the silica precursor.

2.2.3 Condensation

The hydrolyzed silica precursor undergoes a condensation reaction, where the silanol groups react with each other to form siloxane bonds (Si-O-Si) and release water molecules as byproducts. The condensation reaction can be controlled by adjusting reaction parameters such as pH, temperature, and concentration of precursors. The condensation leads to the formation of a three-dimensional silica network.

2.2.4 Pore formation

During the hydrolysis and condensation reactions, the addition of a structure-directing agent or a surfactant can lead to the formation of mesopores within the silica network. The surfactant molecules self-assemble and organize themselves to create micelles, which act as templates for the pore formation. The surfactant can be removed later to generate the mesopores within the MSNs.

2.2.5 Aging and drying

The synthesized MSNs are subjected to aging and drying processes to promote further condensation and solidification. Aging refers to the continuation of the condensation reaction over an extended period, allowing the particles to grow and the pore structure to develop. Drying involves the removal of solvent and water from the MSNs, typically through evaporation or freeze-drying.

The sol-gel method offers several advantages for MSN synthesis. It allows precise control over the particle size, pore size, and pore structure of the MSNs by adjusting reaction parameters. The process is relatively simple, scalable, and compatible with the incorporation of various functional groups or additives for tailored drug delivery applications.

By employing the sol-gel method, researchers can produce MSNs with well-defined mesopores, high surface area, and customizable properties. These MSNs offer advantages for drug delivery, including high drug loading capacity, controlled release profiles, and the potential for targeted delivery to specific tissues or cells. The sol-gel method contributes to the development of efficient and precise drug delivery systems using MSNs as carriers (**Figure 2**).

2.3 Co-condensation method

The co-condensation method is a commonly used technique for the synthesis of mesoporous silica nanoparticles (MSNs) [32–37]. This method involves the simultaneous hydrolysis and condensation of two or more silica precursors, resulting in the formation of MSNs with enhanced structural and chemical properties. The co-condensation method offers advantages such as improved control over pore size, surface functionality, and composition of the MSNs. The following steps outline the co-condensation method for MSN synthesis:

2.3.1 Selection of silica precursors

The co-condensation method utilizes two or more silica precursors, typically alkoxysilanes, such as tetraethyl orthosilicate (TEOS), tetramethyl orthosilicate (TMOS), or organosilanes. The choice of precursors depends on the desired properties and functionalities of the resulting MSNs.

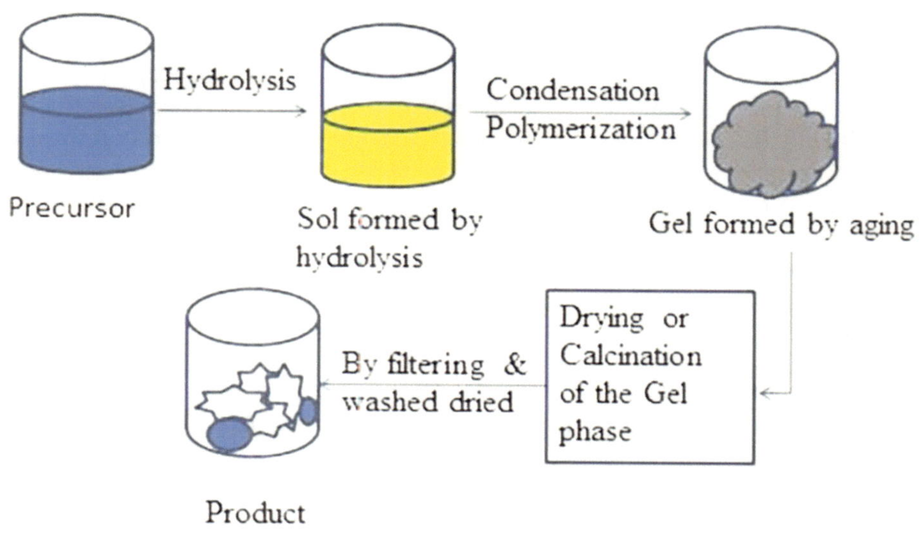

Figure 2.
Sol-gel method for synthesis of mesoporous silica nanoparticles.

2.3.2 Hydrolysis

The selected silica precursors are hydrolyzed by adding water or a hydrolyzing agent, such as an acid or base, to initiate the hydrolysis reaction. Hydrolysis breaks the alkoxide groups in the precursors, leading to the formation of silanol groups (Si-OH) on the silica precursors.

2.3.3 Co-condensation

The hydrolyzed silica precursors undergo co-condensation, where the silanol groups from different precursors react with each other to form siloxane bonds (Si-O-Si). This simultaneous condensation results in the formation of a hybrid silica network containing different types of siloxane linkages.

2.3.4 Pore formation

Similar to other synthesis methods, the addition of structure-directing agents or surfactants can be incorporated during the co-condensation process to induce the formation of mesopores within the silica network. These agents self-assemble and organize themselves to create micelles, which act as templates for pore formation. The subsequent removal of the template generates the desired mesoporous structure within the MSNs.

2.3.5 Aging and drying

The synthesized MSNs undergo aging and drying processes to promote further condensation and solidification. Aging allows the particles to grow and the pore structure to develop over time. Drying involves the removal of solvent and water from the MSNs, typically through evaporation or freeze-drying, resulting in the formation of solid MSNs.

The co-condensation method provides enhanced control over the composition and properties of the MSNs by combining different silica precursors. This method allows the incorporation of organic groups, such as functional molecules or polymers, into the silica network, enabling the introduction of specific functionalities or surface modifications. The resulting MSNs exhibit improved stability, enhanced drug loading capacity, and tailored release profiles, making them suitable for various drug delivery applications.

By utilizing the co-condensation method, researchers can design and synthesize MSNs with specific pore structures, tailored surface functionalities, and controlled drug delivery properties. These MSNs offer advantages in terms of targeted drug delivery, increased stability, and improved therapeutic efficacy. The co-condensation method contributes to the advancement of MSNs as versatile carriers in drug delivery systems (**Figure 3**).

2.4 Other synthesis methods

In addition to template-assisted synthesis and the sol-gel method, there are several other methods available for the synthesis of mesoporous silica nanoparticles (MSNs). These methods offer alternative approaches to tailor the size, shape, and pore characteristics of the MSNs. Here are some of the commonly employed methods:

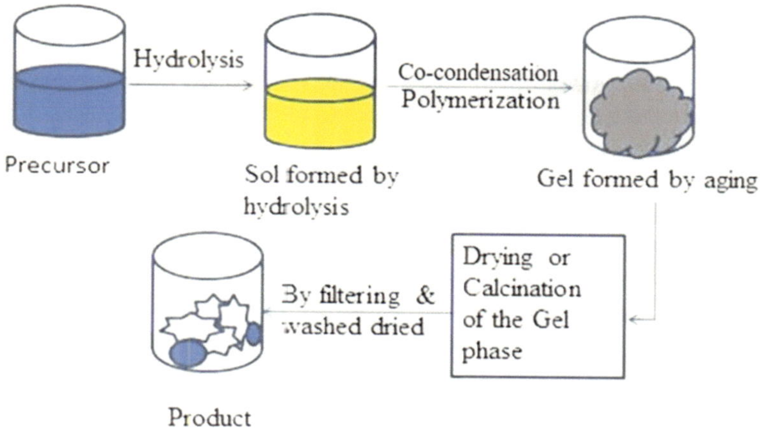

Figure 3.
Co-condensation method for synthesis of mesoporous silica nanoparticles.

2.4.1 Stöber method

The Stöber method, also known as the classical silica nanoparticle synthesis method, involves the hydrolysis and condensation of silica precursors in the presence of a stabilizing agent, such as ammonia or ethanol. The resulting silica nanoparticles can be further etched to generate mesopores, or surfactants can be added during the synthesis process to induce pore formation [38–40].

2.4.2 Microemulsion method

The microemulsion method utilizes water-in-oil or oil-in-water microemulsions as reaction media for the synthesis of MSNs. In this method, silica precursors are dispersed within the microemulsion, and the hydrolysis and condensation reactions take place within the confined nanodroplets. The microemulsion method allows for precise control over the size and morphology of the resulting MSNs [41–44].

2.4.3 Emulsion-droplet coalescence method

The emulsion-droplet coalescence method involves the formation of water-in-oil emulsions containing silica precursors and a hydrophilic solvent. Subsequently, the emulsion droplets are subjected to coalescence, leading to the formation of silica nanoparticles. The resulting nanoparticles can be further treated to introduce mesoporous structures [45–49].

2.4.4 Aerosol-assisted synthesis

The aerosol-assisted synthesis method involves the generation of aerosol droplets containing silica precursors, which are then passed through a high-temperature furnace or reactor. The high temperature promotes the hydrolysis and condensation reactions, resulting in the formation of MSNs. This method allows for the production of MSNs with controlled particle size and narrow size distribution [50, 51].

2.4.5 Solvothermal method

The solvothermal method involves the synthesis of MSNs in a high-pressure, high-temperature environment using organic solvents as reaction media. This method allows for the formation of MSNs with unique structures and properties. The solvothermal conditions facilitate the control of particle size, surface area, and pore characteristics of the MSNs [52].

Each of these alternative methods offers specific advantages in terms of controlling the size, morphology, and pore structure of the resulting MSNs. Researchers can choose the most suitable method based on the desired properties and applications of the nanoparticles. The flexibility and versatility provided by these various synthesis methods contribute to the advancement of MSNs as promising drug delivery systems, enabling tailored approaches to meet specific therapeutic needs.

3. Characterization of MSNs

3.1 Morphological characterization of mesoporous silica nanoparticles

3.1.1 Particle size

Surfactant, which may be charge or neutral, is utilized in aqueous solution to generate mesoporous silica nanoparticles. Surfactant polymerizes silicates, an ester of orthosilicic acid.

The following factors can affect how big and how shaped mesoporous silica nanoparticles are:

- Hydrolysis rate.
- How well the constructed template interacts with the silica polymer.
- Source condensation of silica.

We can modify all of these variables by adjusting the pH and utilizing various templates and co-solvents. Using a high concentration of template and hydrophobic auxiliaries, Stucky et al. created hard mesoporous silica spheres with sizes ranging from hundreds of microns to millimeters at the oil-water interface. The pace of stirring is crucial in determining the particle size of MSNs; if the rate is slow, lengthy fibers are produced, whereas when the rate is fast, one powder is created. Impact of pH on the morphology of MSNs shows that spherical mesoporous particles with a size range of 1–10 m develop in mildly acidic conditions [29].

Although dynamic light scattering is currently favored, electron microscopy was once utilized to measure the particle size of MSNs. MSNs are frequently employed in the biomedical field because, while being smaller than eukaryotic cells, they can function at the subcellular level. MSN interacts with living things like plants, animals, and bacterial cells at both the extracellular and intracellular levels. Spherical MSNs are less likely to be taken up by cancer and non-cancer cells than tubular ones on spherical and tubular MSNs on cancer and non-cancer cells [53].

3.1.2 Pore size

The following variables are employed to regulate the pore shape of MSNs:

- Amount of surfactant and silica source.
- The surfactant's capacity for packing.

Surfactant aggregation in solutions is influenced by pH and solution concentration. Different pore shapes are used to synthesize MSNs at both acidic and basic pH levels. For instance, hexagonal structures are formed at basic pH, but lamellar mesophases are synthesized at high pH (>12). By converting cylindrical channels in an ionic liquid containing MSNs to twisted channels. During or after synthesis, hydrothermal treatment is employed to modify the pore width. In order to produce the required pore size, it is crucial to use a surfactant with varying hydrophobic chain lengths or to use mesitylene as a swelling agent. Since additives have the ability to alter the hydrophobic-hydrophilic equilibrium, further tuning is necessary for pore expansion when applying them during synthesis. Mesitylene is utilized to increase the pore size of MSNs from 3 to 5 nm without changing the particle size, and these MSNs with larger pores are then used as a delivery system for proteins that are membrane impermeable to cancer cells [54]. To enhance pore size without altering the morphology of pre-formed particles, a freshly synthesized material is subjected to autogenic pressure during post-synthesis at temperatures between 373 K and 423 K with or without additions. By combining a fluorocarbon-based and polymer-based surfactant. X-ray diffraction and transmission electron microscopy (TEM) are used to measure the pore structure of MSNs, and nitrogen sorption is used to calculate the pore diameter. The lamellar p2 (MCM-50), the 3D cubic Ia3d, and the 2D hexagonal p6m (MCM-41) are the three most prevalent mesophases in silicas with pore sizes between 2 and 5 nm, respectively. This is comparable to the discovery of large pore size 6–20 nm MSNs with 2D hexagonal p6m [55].

3.1.3 Surface area

The number of medicinal drugs absorbed depends primarily on the surface area of the MSNs. Two distinct methods are utilized to adjust the amount of medication integrated into the matrix: raising or reducing the surface area and changing the surface drug affinity. This shows that the relationship between surface area and drug absorption is direct. Surface area (SBET value) 1157 m^2 g^{-1} and SBA-15 with surface area value 719 m^2 g^{-1} are used to create MCM-41. When alendronate is loaded in MSNs under the identical circumstances, MCM-41 and SBA-15 each receive 139 mg g^{-1} of the medication. It suggests that the relationship between surface area and maximum drug loading is strong [56].

3.1.4 Pore volume

When the surface area is around 1000 ^{2} g^{-1} and the pore size is smaller than 15 nm, the pore volume is typically in the range of 2 cm^2 g^{-1}. Poor drug-drug interactions can cause pore illness while drug interactions with mesopores are a surface phenomenon. The pore volume can be used to calculate the amount of drug adsorbed.

Drug-intermolecular interactions inside the pore width are amplified when drugs are repeatedly loaded into mesopores in ordered mesoporous material. It suggests that the relationship between pore volume and the amount of drug loaded is linear.

Based on the size of their pores, mesoporous materials fall within the category of porous materials. Micro, meso, and microporous are the three categories that can be used to classify the three types of pores. Mesoporous materials typically have pores with a diameter between 2 and 50 nm. Usually, silica is formed around template micelle assemblies to manufacture these materials, and then the templates are removed by calcination. Mesoporous materials' use as drug delivery vehicles has now been expanded due to their distinct pore size, large surface area, and high pore volume. By employing several types of templates and altering the reaction parameters, the pore size of the mesoporous material for such an application can be adjusted. Mesoporous silica nanoparticles (MSN) are stable and expanded mesopores produced by organized mesoporous materials based on silicates produced by sol-gel and hydrothermal synthesis. MSNs are renowned for having excellent characteristics like uniformly shaped pores with well-defined diameters. In addition, altering the template molecule's length can change the pore size of MSNs. By altering the silica sources, surfactants, or reaction parameters, such as temperature, aging time, mole ratio of reactants, and medium pH, a new mesoporous system can also be created [57]. Tetraethyl orthosilicate (TEOS), a precursor to silica, is soluble in alcohol but not in water, hence ethanol is typically used as a homogenizing agent in the reaction. Tetrakis(2-hydroxyethyl) orthosilicate (THEOS), a novel silica precursor with ethylene glycol as a water-soluble residue, was developed to address this problem [58]. THEOS is hydrolyzed into glycol and silicic acid when dissolved in water, and these two substances eventually condense to form silica. The advantage of using THEOS for structured silica fabrication is that it is an environmentally benign technique because the produced silica is readily hydrolyzed and polymerized at neutral pH conditions and can go through jellification at room temperature [59, 60]. Additionally, THEOS as a precursor for silica synthesis only releases glycol rather than alcohols, making it more biocompatible [61].

3.2 Structural characterization of mesoporous silica nanoparticles

Mesoporous nanoparticles have a sizable framework, a porous structure, and a sizable quantity of surface area that allow for the attachment of numerous functional groups for targeting the drug moiety. Chemically, MSNs have a structure resembling a honeycomb and an active surface.

Mesoporous silica nanoparticles (MSNs) combine the benefits of nanomaterials with mesoporous silica materials. This class of materials is characterized by large surface areas, controllable pore size, and ordered pore structures. Particle size, morphology, pore size, and mesostructured can all be controlled through study; a new class of stimuli-responsive aminated MSNs with shape-shifting behavior is introduced. MSNs can change shape by being vacuum-dried from water-rich solvents, being evaporated at high humidity for MSN suspensions in ethanol, or being exposed to water vapor in solid form for 24 h. With the loss of mesostructured long-range hexagonal order, a decrease in surface area and mesopore volume, an increase in micropore volume, and further condensation of the silica matrix, animated MSNs' cross-sectional shapes can change from hexagonal to six-angle stars under these conditions. Finally, a class of quasicrystalline MSNs' synthesis and thorough characterization are addressed. Dodecagonal (12-fold) symmetry is present in these MSNs, which have particles smaller than 100 nm [62].

High surface area, huge pore volume, and consistent and controllable pore size are characteristics of mesoporous silica materials. Because silica is safe, has chemical stability, and can be combined with other materials, it has attracted interest from a variety of fields, including biorelated ones. Nano-sized ordered mesoporous silica particles are one such example.

It has taken a lot of work to create mesoporous silica particles with various shapes, functions, and sizes. On the other hand, the creation mechanism of these particles has received relatively little attention in investigations to date.

The production and characterization of ordered mesoporous silica nanoparticles with and without incorporated magnetic nanoparticles are covered in the first section. By recording particle production at various times during the synthesis, the formation mechanism of silica nano composites is examined. To describe the structure evolution of the resultant materials, transmission electron microscopy (TEM) and small angle x-ray scattering (SAXS) are combined. Ordered mesoporous silica nanoparticles are given additional functionalities by the addition of organic moieties to the silica matrix. However, it frequently results in pore blockage or an unorganized pore structure.

X-ray diffraction (XRD-Pan Analytical X'Pert Pro) was used to determine the crystalline structure, and Fourier transform infrared (FTIR-Thermo Scientific Nicolet IS10) was used to assess the chemical bonding. Microscop electron (SEM-FEI Inspect S50 and TEM) microstructure examination was conducted. Using a Quantochrome surface analyzer and the Brunauer Emmett-Teller (BET) Nitrogen adsorption-desorption method, the specific surface area, pore size, and pore distribution were calculated [63].

4. Surface modification of MSNs

4.1 Functionalization of mesoporous silica nanomaterials

Surface changes have been made for the silica nanoparticles expansion in the bio-domain. By doing so, it is possible to improve biocompatibility, avoid non-specific adsorption, and supply functional groups for future biomolecule conjugation activities. Layer by layer self-assembly (LSA) and chemical surface functionalization are the two most popular surface modifications. The integration of the mesoporous silica nanoparticles during the manufacture of metal or metal oxide nanocrystals can functionalize them. To produce a metal-functionalized silica, a surfactant solution (such as hexadecyltrimethylmonium bromide) must be mixed with a heterogeneous mixture made up of the surfactant-coated metal nanocrystals in an organic solvent. As a result, gold, silver, and iron oxide are embedded into the mesoporous silica. Next, a silicate source is added to the mixture to enhance the condensation reaction. These functionalized systems can exhibit a variety of bio-activities, such as antibacterial activity, which is guaranteed by the presence of dissolved metallic ions [64].

Research has been made to create colloidal core-shell mesoporous silica with various linear PEG (polyethylene glycol) modifications in order to assess the significant impact of various functionalization techniques. Because the PEG matrix is present on the surface of the nanomaterial, silica functionalization can reduce the degradation rate compared to unfunctionalized ones. PEG is hydrophilic, which prevents proteins from adhering to it and minimizes undesirable interactions between the physiological environment and Nano silica. In other studies, silica surface functionalization with hydroxyl, carboxyl, and PEG groups was highlighted. The nanomaterials were largely removed by the renal pathway according to in vivo optical measurements from the urinary bladder, demonstrating that these alterations are independent of the renal clearance. In contrast to the hydroxyl and carboxyl derivates, PEG showed longer blood circulation and reduced liver absorption [65].

The silica functionalization of SBA-15 was the subject of numerous studies. Kim et al. investigated the biodegradation of functionalized SBA-15, modified with hydroxyl, amine, and carboxyl moieties on the surface, in order to assess the impact of surface functionalization onto deterioration behavior. The least amount of degradation was present in the carboxyl functionalized silica. According to these studies, the SBA-15 surface functionalization reduces the degradation rate compared to neat silica, highlighting the possibility that the functionalization may affect silica shell corrosion by interacting with the cations in biological media and thereby slowing the SBA-15 clearance rate. The therapy is typically digested through adsorption, and the mesoporous silica nanoparticles are typically charged by immersion in the active component solution. When the silica surface is functionalized, the cargo can be released in a controlled manner only at the specific damaged tissue; there is no evidence of a premature release into the bloodstream, which minimizes any unwanted side effects and boosts therapeutic effectiveness. The functionalization of SBA-15 mesoporous silica with amino groups from organic amines (aminopropyl triethoxysilane) for the transport of bioactive coordination complexes. The hydrophobic contact with the hydrophobic active principle improved as a result of a linkage between the functional groups from the coordination compound and the amino groups from the silica surface. The rate of medication release will increase once sialylation has been reduced via amination [66].

5. Applications of mesoporous silica nanoparticles (MSNs)

5.1 Drug delivery

Different therapeutic agents, including chemotherapy drugs, small interfering RNA (siRNA), or photo-thermal agents, can be loaded into MSNs. The high drug loading capacity and controlled release made possible by the mesoporous structure increase the efficacy of cancer treatment. Additionally, targeting delivery to particular cancer cells is made possible by functionalizing the MSNs' surface, which minimizes off-target effects [67].

5.2 Imaging

Imaging agents, such as fluorescent dyes, magnetic nanoparticles, or radioactive isotopes, can be added to MSNs. This enables non-invasive imaging of tumor sites, tracking of nanoparticle dispersion, and tracking of therapeutic response. The large surface area of MSNs also enables multi-modal imaging, which combines various imaging modalities for improved diagnostic precision [68].

5.3 Photothermal therapy

MSNs can be used for photothermal therapy by incorporating photothermal agents, such as gold or carbon nano-materials, into them. The photothermal agents produce heat when exposed to near-infrared (NIR) light, selectively ablating cancer cells while sparing healthy tissue. MSNs' mesoporous structure makes it easier to load and deliver photothermal agents to tumor sites effectively [69].

5.4 Combination therapy

MSNs may be made to carry several different therapeutic agents, allowing for the use of combination therapy strategies. Combinations of chemotherapeutic drugs, photothermal agents, immunotherapeutic agents, or gene therapy agents may be used in this. By focusing on multiple pathways involved in cancer progression, these combination therapies have the potential to increase treatment efficacy [70].

5.5 Biosensing and early detection

Functionalized MSNs may be used in biosensing applications to identify particular cancer biomarkers or abnormal cellular processes. These nanoparticles have the ability to bind to target molecules with preference, producing a discernible signal. Early cancer biomarker detection can help with prompt diagnosis and enhance treatment outcomes [71].

5.6 Theranostic nanoplatforms

By combining various functionalities, MSNs can act as adaptable theranostic nanoplatforms. For instance, a single MSN system can be designed to combine photothermal therapy, imaging, and drug delivery capabilities, enabling individualized and accurate cancer treatment.

Although MSNs show great promise in cancer theranostics, more research is still needed to improve their design, increase the effectiveness of their targeting, improve biocompatibility, and guarantee long-term safety. In the coming years, it is anticipated that the clinical translation of MSN-based theranostics will advance as a result of ongoing research in this field [72].

5.7 Catalysis

Mesoporous silica nanoparticles (MSNs) have demonstrated great potential as catalysts in a number of catalytic reactions, including reaction number seven. MSNs are appealing for catalytic applications due to their distinctive structural characteristics, such as their high surface area, large pore volume, and tunable pore size [73–76]. The following are some significant uses of MSNs in catalysis:

5.7.1 Heterogeneous catalysis

To catalyze a variety of reactions, MSNs can be functionalized with different catalytic species, such as metal nanoparticles or metal complexes. Mesoporous channels in MSNs enable effective mass transport of reactants and products, and their large surface area and pore structure provide a sufficient number of active sites for catalytic reactions. In reactions like oxidation, hydrogenation, and selective organic transformations, heterogeneous catalysis using MSNs has been investigated.

5.7.2 Enzyme immobilization

MSNs can effectively support the immobilization of enzymes, enabling the creation of enzyme-based catalysts. Improved stability, reusability, and ease of separation from the reaction mixture are all displayed by enzymes immobilized on MSNs. This makes it possible for enzymatic catalysis in a variety of bio-catalytic processes, such as enzyme-mediated transformations and the creation of biofuels.

5.7.3 Chiral catalysis

To make enantioselective catalysts, MSNs can be functionalized with chiral ligands or chiral metal complexes. These catalysts allow for asymmetric catalysis, which allows for the selective production of particular enantiomers of a compound. Enantiomer separation and effective chiral transformations are made possible by the immobilization of chiral catalysts on MSNs.

5.7.4 Photo-catalysis

To enable photo-catalytic reactions, MSNs can be modified with photo-catalytic species, such as semiconductor nanoparticles (such as $TiO2$, and ZnO). MSNs' mesoporous structure facilitates the interaction of photo-catalysts and reactants and enables effective light harvesting. For uses like water splitting, pollutant degradation, and organic synthesis when exposed to light, photo-catalytic MSNs have been investigated.

5.7.5 Acid-base catalysis

By modifying the surface chemistry of MSNs, acid or base sites can be added, enabling acid-base catalysis. Bronsted or Lewis acid-base sites can be created on the surface of MSNs by grafting functional groups like -SO3H or -NH2. Aldol condensation, trans-esterification, and other reactions have all used MSNs with acid-base functionalities.

The use of MSNs in catalysis offers several advantages, including high catalytic activity, improved selectivity, and recyclability. However, challenges still exist, such as optimizing the catalyst loading, stability under reaction conditions, and mass transfer limitations. Further research is focused on the development of novel MSN-based catalysts and their integration into practical catalytic processes.

6. Toxicity and biocompatibility

6.1 Toxicity

Before using mesoporous silica nanoparticles (MSN) in biomedical applications, it is essential to assess their toxicity [77–79]. MSN's toxicity can change depending on the biological system being studied as well as factors like particle size, surface characteristics, shape, and surface charge. Here are some key points regarding the toxicity of MSN:

6.1.1 Physicochemical characteristics

MSN's toxicity can be greatly influenced by its physicochemical characteristics, such as particle size, surface charge, and surface functionalization. Due to their improved interactions with cellular components, smaller particles and particles with larger surfaces may manifest increased toxicity.

6.1.2 *Inflammatory response*

MSN may cause the release of pro-inflammatory cytokines like interleukin-6 (IL-6) or tumor necrosis factor-alpha (TNF-alpha), which are indicative of an inflammatory response in cells or tissues. Long-lasting or excessive inflammation can have cytotoxic effects and damaged tissue.

6.1.3 *Effects that are dose-dependent*

Like many other nanoparticles, MSN's toxicity frequently shows a dose-dependent relationship. MSN may have negligible or no negative effects at low concentrations, but at higher concentrations, its toxicity may increase.

6.1.4 *Cell type specificity*

Depending on the cell type, exposure to MSN may have different effects. For instance, some studies have indicated that compared to normal cells, some cancer cell lines may be more vulnerable to the toxic effects of MSN. Thus, the toxicity of MSN may vary depending on the type of cell being exposed.

6.1.5 *Cellular uptake and internalization*

MSN can enter cells via a number of different processes, including passive diffusion or endocytosis. The extent of cellular uptake can influence the toxicity of MSN, as internalized nanoparticles may interact with cellular components and induce specific responses.

6.1.6 *Reactive oxygen species (ROS) generation*

MSN can produce ROS that can cause oxidative stress in cells, such as hydrogen peroxide or superoxide radicals. Increased ROS levels have the potential to be cytotoxic by damaging cellular components like proteins, lipids, and DNA.

6.1.7 *Biodistribution and systemic effects*

MSN can interact with different organs and tissues while dispersing throughout the body when taken systemically. When evaluating MSN's toxicity and potential long-term effects on various organ systems, the biodistribution of MSN is a crucial factor to consider.

6.2 Biocompatibility

6.2.1 *IN-VIVO studies*

Due to their potential use in drug delivery, imaging, and diagnostics, mesoporous silica nanoparticles (MSN) have drawn a lot of interest in biomedical research. Before MSN is put to use in clinical settings, biocompatibility studies are essential for determining its safety and effectiveness. MSN interactions with living organisms are examined in vivo studies, particularly in terms of biocompatibility, bio-distribution and potential toxicity [80–83].

Animal models like mice, rats, or non-human primates are frequently given these nanoparticles in biocompatibility studies of MSN. Depending on the intended use, the nanoparticles can be given through a variety of routes, such as intravenous injection, oral administration, or inhalation. The animals are then observed for a predetermined amount of time to evaluate any negative effects or biological reactions.

To evaluate the biocompatibility of MSN, several aspects are typically investigated:

6.2.1.1 Acute and chronic toxicity

Animals are watched closely for any indications of acute toxicity, such as modifications in behavior, body weight, or organ function. Studies on chronic toxicity examine the long-term consequences of MSN exposure, such as possible organ accumulation, inflammation, or organ dysfunction.

6.2.1.2 Bio-distribution

To comprehend the uptake, localization, and clearance mechanisms of MSN, the distribution of MSN in various organs and tissues is examined. The distribution of nanoparticles can be seen and measured using methods like fluorescence imaging, electron microscopy, and radiolabeling.

6.2.1.3 Immunological response

To determine any potential immune toxicity, the immune response induced by MSN is examined. To ascertain whether MSN causes an inflammatory response or immune cell activation, immune cells, cytokine levels, and histological analysis of immune-related organs are examined.

6.2.1.4 Organ-specific effects

To assess any potential harm or dysfunction brought on by MSN, specific organ systems, such as the liver, kidneys, lungs, and spleen, are carefully examined. It is common practice to evaluate tissue morphology and spot any anomalies through histopathological analysis.

6.2.1.5 Metabolism and excretion

To comprehend the biotransformation and excretion pathways of MSN, the metabolic fate of the substance is investigated. To determine their potential toxicity, metabolites and degradation products are examined.

6.2.2 IN-VITRO studies

For mesoporous silica nanoparticles (MSN), in-vitro studies are a crucial part of biocompatibility evaluations. Insights into their potential toxicity, cellular uptake mechanisms, and biological responses can be gained from these studies by analyzing the interactions between MSN and various biological components at the cellular and molecular levels [84–89].

Here are a few typical in vitro techniques for evaluating MSN's biocompatibility:

Tests for cytotoxicity and cell viability: These tests examine the viability, proliferation, and metabolic activity of cultured cells after they have been exposed to MSN. The MTT assay, the Cell Counting Kit-8 (CCK-8) assay, or the lactate dehydrogenase (LDH) release assay are examples of frequently used assays. Potential side effects may be indicated by any appreciable decrease in cell viability or rise in cytotoxicity when compared to control groups.

6.2.2.1 Cellular uptake and internalization

In vitro studies investigate the interactions between MSN and cells, including the internalization processes and intracellular fate of MSN. To see and measure the uptake of MSN by cells, techniques like fluorescence microscopy, flow cytometry, or electron microscopy can be used.

6.2.2.2 Inflammatory response

The ability of MSN to cause an inflammatory response is determined by monitoring the release of pro-inflammatory cytokines by immune cells or other pertinent cell types, such as interleukin-6 (IL-6) or tumor necrosis factor-alpha (TNF-alpha). To measure the expression of cytokines, real-time PCR or enzyme-linked immunosorbent assays (ELISA) are frequently used.

6.2.2.3 Oxidative stress assessment

MSN may produce reactive oxygen species (ROS) inside of cells, which causes oxidative stress. Studies conducted in vitro measure oxidative stress markers like lipid peroxidation, superoxide dismutase activity, intracellular ROS levels, and other antioxidant enzyme activities. These analyses aid in assessing the possibility of oxidative damage brought on by MSN.

6.2.2.4 Genotoxicity and DNA damage

In vitro genotoxicity assays, such as the micronucleus assay or the comet assay, assess the likelihood that MSN will result in genomic instability or DNA damage in cells. The results of these tests shed light on the potential long-term consequences of MSN exposure.

6.2.2.5 Cell-specific assays

Additional cell-specific assays may be carried out depending on the intended use of MSN. Studies may assess the release of encapsulated drugs, therapeutic efficacy, or particular cellular reactions to the delivered cargo, for instance, if MSN is intended for drug delivery.

7. Conclusions

MSNs offer unique properties that make them highly attractive for drug delivery systems. Their high surface area, tunable pore size, and excellent biocompatibility make them suitable for efficient encapsulation, controlled release, and targeted delivery of therapeutic agents.

The chapter discusses various synthesis methods for MSNs, including template-assisted synthesis, sol-gel method, co-condensation method, and other approaches. Each method offers specific advantages and allows for the customization of MSNs with desired characteristics. The characterization techniques for evaluating MSNs, such as morphological, structural, and chemical characterization, are also presented, emphasizing the importance of assessing the quality and functionality of these nanoparticles.

Surface modification of MSNs is explored, highlighting the strategies for functionalizing surface groups, attaching targeting ligands, and modifying surface charge. These modifications enable improved interactions with specific cells or tissues, enhancing the efficacy and specificity of drug delivery.

The chapter further discusses the diverse applications of MSNs, focusing on cancer theranostics, drug delivery, imaging, biosensing, and catalysis. MSNs show great potential in revolutionizing these areas by enabling precise drug delivery, multimodal imaging, sensitive biosensing, and efficient catalytic reactions.

Toxicity and biocompatibility of MSNs are addressed, covering in vitro and in vivo studies that evaluate the safety and efficacy of these nanoparticles. The understanding of their biocompatibility is essential for their successful translation into clinical applications.

Finally, the chapter concludes by highlighting future research directions in the field of MSNs. Ongoing research aims to improve the design and fabrication of MSNs, enhance their drug loading and release capabilities, explore new applications, and address any potential challenges related to toxicity and biocompatibility. The significant potential of MSNs in advancing drug delivery systems is underscored, emphasizing their role in the development of innovative and targeted therapeutic strategies.

Acknowledgements

The authors would like to express their sincere gratitude to the Management of RSM's N. N. Sattha College of Pharmacy, Ahmednagar, for providing the necessary resources and support for the completion of this work. We would also like to acknowledge the valuable contributions of our colleagues who provided insightful discussions and suggestions throughout the research process.

Conflict of interest

The authors declare no conflict of interest.

Author details

Vishal Pande[1]*, Sachin Kothawade[1], Sharmila Kuskar[1], Sandesh Bole[1] and Dinesh Chakole[2]

1 RSM's N.N. Sattha College of Pharmacy, Ahmednagar, MH, India

2 Pacific Academy of Higher Education and Research University, Udaipur, India

*Address all correspondence to: drvishalpande@gmail.com

References

[1] Porrang S, Davaran S, Rahemi N, Allahyari S, Mostafavi E. How advancing are mesoporous silica nanoparticles? A comprehensive review of the literature. International Journal of Nanomedicine. 2022;**2022**:1803-1827. DOI: 10.2147/IJN.S353349

[2] Pu X, Li J, Qiao P, Li M, Wang H, Zong L, et al. Mesoporous silica nanoparticles as a prospective and promising approach for drug delivery and biomedical applications. Current Cancer Drug Targets. 2019;**19**(4):285-295. DOI: 10.2174/1568009619666181206114904

[3] Rey-Raap N, Menéndez JA, Arenillas A. RF xerogels with tailored porosity over the entire nanoscale. Microporous and Mesoporous Materials. 2014;**195**:266-275. DOI: 10.1016/j.micromeso.2014.04.048

[4] Bagheri E, Ansari L, Abnous K, Taghdisi SM, Charbgoo F, Ramezani M, et al. Silica based hybrid materials for drug delivery and bioimaging. Journal of Controlled Release. 2018;**277**:57-76. DOI: 10.1016/j.jconrel.2018.03.014

[5] Chen W, Glackin CA, Horwitz MA, Zink JI. Nanomachines and other caps on mesoporous silica nanoparticles for drug delivery. Accounts of Chemical Research. 2019;**52**(6):1531-1542. DOI: 10.1021/acs.accounts.9b00116

[6] Alyassin Y, Sayed EG, Mehta P, Ruparelia K, Arshad MS, Rasekh M, et al. Application of mesoporous silica nanoparticles as drug delivery carriers for chemotherapeutic agents. Drug Discovery Today. 2020;**25**(8):1513-1520. DOI: 10.1016/j.drudis.2020.06.006

[7] Stephen S, Gorain B, Choudhury H, Chatterjee B. Exploring the role of mesoporous silica nanoparticle in the development of novel drug delivery systems. Drug Delivery and Translational Research. 2021;**2021**:105-123. DOI: 10.1007/s13346-021-00935-4

[8] Radhakrishnan D, Mohanan S, Choi G, Choy JH, Tiburcius S, Trinh HT, et al. The emergence of nanoporous materials in lung cancer therapy. Science and Technology of Advanced Materials. 2022;**23**(1):225-274. DOI: 10.1080/14686996.2022.2052181

[9] Singh LP, Bhattacharyya SK, Kumar R, Mishra G, Sharma U, Singh G, et al. Sol-gel processing of silica nanoparticles and their applications. Advances in Colloid and Interface Science. 2014;**214**:17-37. DOI: 10.1016/j.cis.2014.10.007

[10] Long R, Zhou S, Wiley BJ, Xiong Y. Oxidative etching for controlled synthesis of metal nanocrystals: Atomic addition and subtraction. Chemical Society Reviews. 2014;**43**(17):6288-6310. DOI: 10.1039/C5SC01947H

[11] Li H, Chen X, Shen D, Wu F, Pleixats R, Pan J. Functionalized silica nanoparticles: Classification, synthetic approaches and recent advances in adsorption applications. Nanoscale. 2021;**13**(38):15998-16016. DOI: 10.1039/D1NR04048K

[12] Agrahari V, Burnouf PA, Burnouf T, Agrahari V. Nanoformulation properties, characterization, and behavior in complex biological matrices: Challenges and opportunities for brain-targeted drug delivery applications and enhanced translational potential. Advanced Drug Delivery Reviews. 2019;**148**:146-180. DOI: 10.1016/j.addr.2019.02.008

[13] Agrawal G, Agrawal R. Janus nanoparticles: Recent advances in their interfacial and biomedical applications. ACS Applied Nano Materials. 2019;**2**(4):1738-1757. DOI: 10.1021/acsanm.9b00283

[14] Zhu H, Zheng K, Boccaccini AR. Multi-functional silica-based mesoporous materials for simultaneous delivery of biologically active ions and therapeutic biomolecules. Acta Biomaterialia. 2021;**129**:1-17. DOI: 10.1016/j.actbio.2021.05.007

[15] Wuttke S, Lismont M, Escudero A, Rungtaweevoranit B, Parak WJ. Positioning metal-organic framework nanoparticles within the context of drug delivery–a comparison with mesoporous silica nanoparticles and dendrimers. Biomaterials. 2017;**123**:172-183. DOI: 10.1016/j.biomaterials.2017.01.025

[16] Croissant JG, Butler KS, Zink JI, Brinker CJ. Synthetic amorphous silica nanoparticles: Toxicity, biomedical and environmental implications. Nature Reviews Materials. 2020;**5**(12):886-909. DOI: 10.1038/s41578-020-0230-0

[17] Maity A, Belgamwar R, Polshettiwar V. Facile synthesis to tune size, textural properties and fiber density of dendritic fibrous nanosilica for applications in catalysis and CO_2 capture. Nature Protocols. 2019;**14**(7):2177-2204. DOI: 10.1038/s41596-019-0177-z

[18] Lee EC, Nguyen CT, Strounina E, Davis-Poynter N, Ross BP. Structure–activity relationships of GAG mimetic-functionalized mesoporous silica nanoparticles and evaluation of acyclovir-loaded antiviral nanoparticles with dual mechanisms of action. ACS Omega. 2018;**3**(2):1689-1699. DOI: 10.1021/acsomega.7b01662

[19] Živojević K, Mladenović M, Djisalov M, Mundzic M, Ruiz-Hernandez E, Gadjanski I, et al. Advanced mesoporous silica nanocarriers in cancer theranostics and gene editing applications. Journal of Controlled Release. 2021;**337**:193-211. DOI: 10.1016/j.jconrel.2021.07.029

[20] Bikiaris DN. Solid dispersions, part II: New strategies in manufacturing methods for dissolution rate enhancement of poorly water-soluble drugs. Expert Opinion on Drug Delivery. 2011;**8**(12):1663-1680. DOI: 10.1517/17425247.2011.618182

[21] Slowing II, Vivero-Escoto JL, Wu CW, Lin VS. Mesoporous silica nanoparticles as controlled release drug delivery and gene transfection carriers. Advanced Drug Delivery Reviews. 2008;**60**(11):1278-1288. DOI: 10.1016/j.addr.2008.03.012

[22] He Q, Shi J. Mesoporous silica nanoparticle based nano drug delivery systems: Synthesis, controlled drug release and delivery, pharmacokinetics and biocompatibility. Journal of Materials Chemistry. 2011;**21**(16):5845-5855. DOI: 10.1039/C0JM03851B

[23] Wu KC, Yamauchi Y. Controlling physical features of mesoporous silica nanoparticles (MSNs) for emerging applications. Journal of Materials Chemistry. 2012;**22**(4):1251-1256. DOI: 10.1039/C1JM13811A

[24] Kong XP, Zhang BH, Wang J. Multiple roles of mesoporous silica in safe pesticide application by nanotechnology: A review. Journal of Agricultural and Food Chemistry. 2021;**69**(24):6735-6754. DOI: 10.1021/acs.jafc.1c01091

[25] Pérez-Page M, Yu E, Li J, Rahman M, Dryden DM, Vidu R, et al. Template-based syntheses for shape-controlled nanostructures. Advances in Colloid and Interface Science. 2016;**234**:51-79. DOI: 10.1016/j.cis.2016.04.001

[26] Zuo B, Li W, Wu X, Wang S, Deng Q, Huang M. Recent advances in the synthesis, surface modifications and applications of core-shell magnetic mesoporous silica nanospheres. Chemistry–An Asian Journal. 2020;**15**(8):1248-1265. DOI: 10.1002/asia.202000045

[27] Trewyn BG, Slowing II, Giri S, Chen HT, Lin VS. Synthesis and functionalization of a mesoporous silica nanoparticle based on the sol–gel process and applications in controlled release. Accounts of Chemical Research. 2007;**40**(9):846-853. DOI: 10.1021/ar600032u

[28] Kumar S, Malik MM, Purohit R. Synthesis methods of mesoporous silica materials. Materials Today: Proceedings. 2017;**4**(2):350-357. DOI: 10.1016/j.matpr.2017.01.032

[29] Wu SH, Mou CY, Lin HP. Synthesis of mesoporous silica nanoparticles. Chemical Society Reviews. 2013;**42**(9):3862-3875. DOI: 10.1039/C3CS35405A

[30] Chang B, Guo J, Liu C, Qian J, Yang W. Surface functionalization of magnetic mesoporous silica nanoparticles for controlled drug release. Journal of Materials Chemistry. 2010;**20**(44):9941-9947. DOI: 10.1039/C0JM01237H

[31] Kajani AA, Rafiee L, Javanmard SH, Dana N, Jandaghian S. Carbon dot incorporated mesoporous silica nanoparticles for targeted cancer therapy and fluorescence imaging. RSC Advances. 2023;**13**(14):9491-9500. DOI: 10.1039/D3RA00768E

[32] Kamarudin NH, Jalil AA, Triwahyono S, Salleh NF, Karim AH, Mukti RR, et al. Role of 3-aminopropyltriethoxysilane in the preparation of mesoporous silica nanoparticles for ibuprofen delivery: Effect on physicochemical properties. Microporous and Mesoporous Materials. 2013;**180**:235-241. DOI: 10.1016/j.micromeso.2013.06.041

[33] Möller K, Bein T. Talented mesoporous silica nanoparticles. Chemistry of Materials. 2017;**29**(1):371-388. DOI: 10.1021/acs.chemmater.6b03629

[34] von Baeckmann C, Guillet-Nicolas R, Renfer D, Kählig H, Kleitz F. A toolbox for the synthesis of multifunctionalized mesoporous silica nanoparticles for biomedical applications. ACS Omega. 2018;**3**(12):17496-17510. DOI: 10.1021/acsomega.8b02784

[35] Estevão BM, Miletto I, Hioka N, Marchese L, Gianotti E. Mesoporous silica nanoparticles functionalized with amino groups for biomedical applications. ChemistryOpen. 2021;**10**(12):1251-1259. DOI: 10.1002/open.202100227

[36] Chiu HY, Gößl D, Haddick L, Engelke H, Bein T. Clickable multifunctional large-pore mesoporous silica nanoparticles as nanocarriers. Chemistry of Materials. 2018;**30**(3):644-654. DOI: 10.1021/acs.chemmater.7b03472

[37] Talavera-Pech WA, Ávila-Ortega A, Pacheco-Catalán D, Quintana-Owen P, Barrón-Zambrano JA. Effect of functionalization synthesis type of amino-MCM-41 mesoporous silica nanoparticles on its RB5 adsorption capacity and kinetics. Silicon. 2019;**11**:1547-1555. DOI: 10.1007/s12633-018-9975-0

[38] Semeykina V, Zharov I. Medium controlled aggregative growth as a key step in mesoporous silica nanoparticle formation. Journal of Colloid and Interface Science. 2022;**615**:236-247. DOI: 10.1016/j.jcis.2022.01.166

[39] Lin CH, Chang JH, Yeh YQ, Wu SH, Liu YH, Mou CY. Formation of hollow silica nanospheres by reverse microemulsion. Nanoscale. 2015;**7**(21):9614-9626. DOI: 10.1039/C5NR01395J

[40] Singh B, Na J, Konarova M, Wakihara T, Yamauchi Y, Salomon C, et al. Functional mesoporous silica nanomaterials for catalysis and environmental applications. Bulletin of the Chemical Society of Japan. 2020;**93**(12):1459-1496. DOI: 10.1246/bcsj.20200136

[41] Liu J, Qiao SZ, Hu QH, Lu GQ. Magnetic nanocomposites with mesoporous structures: Synthesis and applications. Small. 2011;7(4):425-443. DOI: 10.1002/smll.201001402

[42] Douroumis D, Onyesom I, Maniruzzaman M, Mitchell J. Mesoporous silica nanoparticles in nanotechnology. Critical Reviews in Biotechnology. 2013;**33**(3):229-245. DOI: 10.3109/07388551.2012.685860

[43] Thakur N, Manna P, Das J. Synthesis and biomedical applications of nanoceria, a redox active nanoparticle. Journal of Nanobiotechnology. 2019;**17**(1):1-27. DOI: 10.1186/s12951-019-0516-9

[44] Maleki A, Kettiger H, Schoubben A, Rosenholm JM, Ambrogi V, Hamidi M. Mesoporous silica materials: From physico-chemical properties to enhanced dissolution of poorly water-soluble drugs. Journal of Controlled Release. 2017;**262**:329-347. DOI: 10.1016/j.jconrel.2017.07.047

[45] Akartuna I, Studart AR, Tervoort E, Gauckler LJ. Macroporous ceramics from particle-stabilized emulsions. Advanced Materials. 2008;**20**(24):4714-4718. DOI: 10.1002/adma.200801888

[46] Medina-Sandoval CF, Valencia-Davila JA, Combariza MY, Blanco-Tirado C. Separation of asphaltene-stabilized water in oil emulsions and immiscible oil/water mixtures using a hydrophobic cellulosic membrane. Fuel. 2018;**231**:297-306. DOI: 10.1016/j.fuel.2018.05.066

[47] Perro A, Coudon N, Chapel JP, Martin N, Béven L, Douliez JP. Building micro-capsules using water-in-water emulsion droplets as templates. Journal of Colloid and Interface Science. 2022;**613**:681-696. DOI: 10.1016/j.jcis.2022.01.047

[48] Sheth T, Seshadri S, Prileszky T, Helgeson ME. Multiple nanoemulsions. Nature Reviews Materials. 2020;**5**(3):214-228. DOI: 10.1038/s41578-019-0161-9

[49] Thompson KL, Mable CJ, Lane JA, Derry MJ, Fielding LA, Armes SP. Preparation of Pickering double emulsions using block copolymer worms. Langmuir. 2015;**31**(14):4137-4144. DOI: 10.1021/acs.langmuir.5b00741

[50] Bang JH, Suslick KS. Applications of ultrasound to the synthesis of nanostructured materials. Advanced Materials. 2010;**22**(10):1039-1059. DOI: 10.1002/adma.200904093

[51] Yang Y, Song B, Ke X, Xu F, Bozhilov KN, Hu L, et al. Aerosol synthesis of high entropy alloy nanoparticles. Langmuir. 2020;**36**(8):1985-1992. DOI: 10.1021/acs.langmuir.9b03392

[52] Sharma N, Ojha H, Bharadwaj A, Pathak DP, Sharma RK. Preparation and catalytic applications of nanomaterials: A review. RSC Advances. 2015;**5**(66):53381-53403. DOI: 10.1039/C5RA06778B

[53] Zare I, Yaraki MT, Speranza G, Najafabadi AH, Shourangiz-Haghighi A,

Nik AB, et al. Gold nanostructures: Synthesis, properties, and neurological applications. Chemical Society Reviews. 2022;**51**(7):2601-2680. DOI: 10.1039/D1CS01111A

[54] Melero JA, Iglesias J, Morales G. Designing porous inorganic architectures. Heterogeneous Catalysts for Clean Technology: Spectroscopy, Design, and Monitoring. 2013;**23**:193-240. DOI: 10.1002/9783527658985.ch8 3

[55] Wu L, Jiao Z, Wu M, Song T, Zhang H. Formation of mesoporous silica nanoparticles with tunable pore structure as promising nanoreactor and drug delivery vehicle. RSC Advances. 2016;**6**(16):13303-13311. DOI: 10.1039/C5RA27422B

[56] Popat A, Hartono SB, Stahr F, Liu J, Qiao SZ, Lu GQ. Mesoporous silica nanoparticles for bioadsorption, enzyme immobilisation, and delivery carriers. Nanoscale. 2011;**3**(7):2801-2818. DOI: 10.1039/C1NR10224A

[57] ALOthman ZA. A review: Fundamental aspects of silicate mesoporous materials. Materials. 2012;**5**(12):2874-2902. DOI: 10.3390/ma5122874

[58] Boissiere C, Larbot A, van der Lee A, Kooyman PJ, Prouzet E. A new synthesis of mesoporous MSU-X silica controlled by a two-step pathway. Chemistry of Materials. 2000;**12**(10):2902-2913. DOI: 10.1021/cm991188s

[59] O'Brien S, Francis RJ, Fogg A, O'Hare D, Okazaki N, Kuroda K. Time-resolved in situ X-ray powder diffraction study of the formation of mesoporous silicates. Chemistry of Materials. 1999;**11**(7):1822-1832. DOI: 10.1021/cm990044a

[60] Owens GJ, Singh RK, Foroutan F, Alqaysi M, Han CM, Mahapatra C, et al. Sol–gel based materials for biomedical applications. Progress in Materials Science. 2016;**77**:1-79. DOI: 10.1016/j.pmatsci.2015.12.001

[61] Morales V, McConnell J, Pérez-Garnes M, Almendro N, Sanz R, García-Muñoz RA. l-Dopa release from mesoporous silica nanoparticles engineered through the concept of drug-structure-directing agents for Parkinson's disease. Journal of Materials Chemistry B. 2021;**9**(20):4178-4189. DOI: 10.1039/D1TB00481F

[62] Kolimi P, Narala S, Youssef AA, Nyavanandi D, Dudhipala N. A systemic review on development of mesoporous nanoparticles as a vehicle for transdermal drug delivery. Nano. 2023;**7**(1):70-89. DOI: 10.7150/ntno.77395

[63] Yorgun S, Vural N, Demiral H. Preparation of high-surface area activated carbons from Paulownia wood by $ZnCl_2$ activation. Microporous and Mesoporous Materials. 2009;**122**(1-3):189-194. DOI: 10.1016/j.micromeso.2009.02.032

[64] Musa MA, Yin CY, Savory RM. Analysis of the textural characteristics and pore size distribution of a commercial zeolite using various adsorption models. Journal of Applied Sciences. 2011;**11**(21):3650-3654. DOI: 10.3923/jas.2011.3650.3654

[65] Wang S, Wang L, Yang T, Liu X, Zhang J, Zhu B, et al. Porous α-Fe2O3 hollow microspheres and their application for acetone sensor. Journal of Solid State Chemistry. 2010;**183**(12):2869-2876. DOI: 10.1016/j.jssc.2010.09.033

[66] Xu MW, Zhao DD, Bao SJ, Li HL. Mesoporous amorphous MnO_2 as electrode material for supercapacitor. Journal of Solid State Electrochemistry.

2007;**11**:1101-1107. DOI: 10.1007/s10008-006-0246-4

[67] Liu F, Wang C, Gao Y, Li X, Tian F, Zhang Y, et al. Current transport systems and clinical applications for small interfering RNA (siRNA) drugs. Molecular Diagnosis & Therapy. 2018;**22**:551-569. DOI: 10.1007/s40291-018-0338-8

[68] Lee N, Hyeon T. Designed synthesis of uniformly sized iron oxide nanoparticles for efficient magnetic resonance imaging contrast agents. Chemical Society Reviews. 2012;**41**(7):2575-2589. DOI: 10.1039/C1CS15248C

[69] Zhi D, Yang T, O'hagan J, Zhang S, Donnelly RF. Photothermal therapy. Journal of Controlled Release. 2020;**325**:52-71. DOI: 10.1016/j.jconrel.2020.06.032

[70] Qin SY, Cheng YJ, Lei Q, Zhang AQ, Zhang XZ. Combinational strategy for high-performance cancer chemotherapy. Biomaterials. 2018;**171**:178-197. DOI: 10.1016/j.biomaterials.2018.04.027

[71] Toy R, Bauer L, Hoimes C, Ghaghada KB, Karathanasis E. Targeted nanotechnology for cancer imaging. Advanced Drug Delivery Reviews. 2014;**76**:79-97. DOI: 10.1016/j.addr.2014.08.002

[72] Yang B, Chen Y, Shi J. Mesoporous silica/organosilica nanoparticles: Synthesis, biological effect and biomedical application. Materials Science and Engineering: R: Reports. 2019;**137**:66-105. DOI: 10.1016/j.mser.2019.01.001

[73] Sharma RK, Sharma S, Dutta S, Zboril R, Gawande MB. Silica-nanosphere-based organic–inorganic hybrid nanomaterials: Synthesis, functionalization and applications in catalysis. Green Chemistry. 2015;**17**(6):3207-3230. DOI: 10.1039/C5GC00381D

[74] Sharma RK, Dutta S, Sharma S, Zboril R, Varma RS, Gawande MB. Fe_3O_4 (iron oxide)-supported nanocatalysts: Synthesis, characterization and applications in coupling reactions. Green Chemistry. 2016;**18**(11):3184-3209. DOI: 10.1039/C6GC00864J

[75] Bhanja P, Modak A, Bhaumik A. Supported porous nanomaterials as efficient heterogeneous catalysts for CO_2 fixation reactions. Chemistry: A European Journal. 2018;**24**(29):7278-7297. DOI: 10.1002/chem.201800075

[76] Shegavi ML, Bose SK. Recent advances in the catalytic hydroboration of carbonyl compounds. Catalysis Science & Technology. 2019;**9**(13):3307-3336. DOI: 10.1039/C9CY00807A

[77] Hosseinpour S, Walsh LJ, Xu C. Biomedical application of mesoporous silica nanoparticles as delivery systems: A biological safety perspective. Journal of Materials Chemistry B. 2020;**8**(43):9863-9876. DOI: 10.1039/D0TB01868F

[78] Asefa T, Tao Z. Biocompatibility of mesoporous silica nanoparticles. Chemical Research in Toxicology. 2012;**25**(11):2265-2284. DOI: 10.1021/tx300166u

[79] Hossen S, Hossain MK, Basher MK, Mia MN, Rahman MT, Uddin MJ. Smart nanocarrier-based drug delivery systems for cancer therapy and toxicity studies: A review. Journal of Advanced Research. 2019;**15**:1-8. DOI: 10.1016/j.jare.2018.06.005

[80] Chen KJ, Plaunt AJ, Leifer FG, Kang JY, Cipolla D. Recent advances in prodrug-based nanoparticle therapeutics.

European Journal of Pharmaceutics and Biopharmaceutics. 2021;**165**:219-243. DOI: 10.1016/j.ejpb.2021.04.025

[81] Attia MF, Anton N, Wallyn J, Omran Z, Vandamme TF. An overview of active and passive targeting strategies to improve the nanocarriers efficiency to tumour sites. Journal of Pharmacy and Pharmacology. 2019;**71**(8):1185-1198. DOI: 10.1111/jphp.13098

[82] Kumar R, Mondal K, Panda PK, Kaushik A, Abolhassani R, Ahuja R, et al. Core–shell nanostructures: Perspectives towards drug delivery applications. Journal of Materials Chemistry B. 2020;**8**(39):8992-9027. DOI: 10.1039/D0TB01559H

[83] Deng F, Bae YH. Bile acid transporter-mediated oral drug delivery. Journal of Controlled Release. 2020;**327**:100-116. DOI: 10.1016/j.jconrel.2020.07.034

[84] Tang F, Li L, Chen D. Mesoporous silica nanoparticles: Synthesis, biocompatibility and drug delivery. Advanced Materials. 2012;**24**(12):1504-1534. DOI: 10.1002/adma.201104763

[85] Vivero-Escoto JL, Slowing II, Trewyn BG, Lin VS. Mesoporous silica nanoparticles for intracellular controlled drug delivery. Small. 2010;**6**(18):1952-1967. DOI: 10.1002/smll.200901789

[86] Argyo C, Weiss V, Bräuchle C, Bein T. Multifunctional mesoporous silica nanoparticles as a universal platform for drug delivery. Chemistry of Materials. 2014;**26**(1):435-451. DOI: 10.1021/cm402592t

[87] Hao N, Li L, Tang F. Shape-mediated biological effects of mesoporous silica nanoparticles. Journal of Biomedical Nanotechnology. 2014;**10, 10**:2508-2538. DOI: 10.1166/jbn.2014.1940

[88] Hu B, Wang J, Li J, Li S, Li H. Superiority of L-tartaric acid modified chiral mesoporous silica nanoparticle as a drug carrier: Structure, wettability, degradation, bio-adhesion and biocompatibility. International Journal of Nanomedicine. 2020;**29**:601-618. DOI: 10.2147/IJN.S233740

[89] Li H, Wu X, Yang B, Li J, Xu L, Liu H, et al. Evaluation of biomimetically synthesized mesoporous silica nanoparticles as drug carriers: Structure, wettability, degradation, biocompatibility and brain distribution. Materials Science and Engineering: C. 2019;**94**:453-464. DOI: 10.1016/j.msec.2018.09.053

Chapter 4

ZnO Nanofiber Fabrication Technique

Thangavel Krishnasamy, Balaprakash Vadivel, Prema Rangasamy and Mahitha Mohan

Abstract

Zinc oxide (ZnO) nanofibers have gained significant attention due to their unique properties and potential applications in various fields, including electronics, optoelectronics, sensors, energy storage, and biomedical devices. This chapter presents an overview of different fabrication techniques employed for the synthesis of ZnO nanofibers. It discusses both template-based and template-free methods, highlighting their advantages, limitations, and the resulting morphological and structural characteristics of the fabricated nanofibers. Electrospinning is a crucial nanofiber fabrication technique utilized in electronics design and software to create high-performance materials with tailored properties for applications such as sensors, energy storage devices, and electronic components. Furthermore, the chapter provides insights into the influence of process parameters on the growth mechanism and properties of ZnO nanofibers. The goal is to provide readers with a comprehensive understanding of the various techniques available for fabricating ZnO nanofibers and to guide them in selecting an appropriate method for their specific applications.

Keywords: ZnO nanofibers, fabrication techniques, template-based methods, template-free methods, growth mechanism, properties

1. Introduction

1.1 Overview of ZnO nanofibers

"There's plenty of room at the bottom", said Richard Feynman in 1959. The fabrication and characterization of such rooms (nano structures) have imbibed in the mind of each and every researchers today. Nanotechnology has emerged as an exciting research field in the last decade. Recently, nanostructures have attracted extensive research interests because of their unique properties. They have made innovations in nanotechnology, particularly nano sensor devices. Nano structures include nano particles, nano films, nano ribbon, nano flower, nano cluster, nanofibers, etc., all of them are frontier materials and gain dominant importance in nanotechnology. **Figure 1** shows the various nano and microscale illustrations.

Currently, nanofibers occupy a prominent position in the realm of nanotechnology due to their exceptional attributes encompassing reduced density, exceedingly

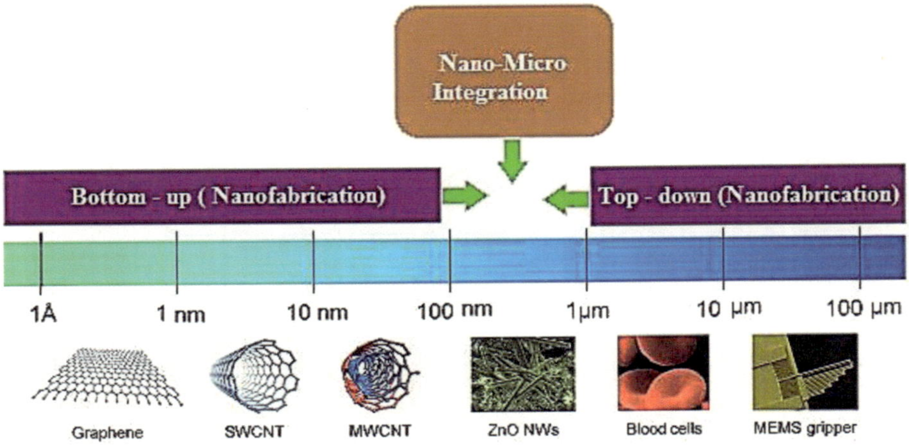

Figure 1.
Nano-micro scale illustrations.

high surface area in relation to weight, significant pore volume, and adjustable pore size. These distinctive characteristics render nonwoven nanofibers highly suitable for a wide array of applications. Moreover, these nanofibers exhibit the ability to imitate the structure, chemical composition, and other properties of fiber matrices. Consequently, nanofibers play a pivotal role in various fields such as biomedicine, skin tissue engineering, bone regeneration, wound healing, vascular grafts, and drug delivery systems.

Metal oxide nanofibers are attractive components for nanometer scale electronic devices like LED, Photodiodes, varistors, solar cells, etc., the metal oxides ZnO, SnO_2 and CuO have exceptional blend of interesting properties such as non-toxicity, good electrical properties, high luminous transmittance, excellent substrate adherence, hardness, optical and piezoelectric behaviors, stability in atmosphere and it low cost.

Electrospinning emerges as the most promising technique in the realm of nanotechnology, primarily owing to its simplicity, cost-effectiveness, high productivity, reproducibility, and potential for industrial scalability. This method involves applying a high voltage electric field to extract ultra-thin fibers from a stream of polymeric fluid (solution or melt) delivered through a nozzle measuring in millimeters [1]. The electrospinning process relies on several processing parameters, including solution properties (such as viscosity, surface tension, and conductivity) and other factors (like electric field strength, solution flow rate, needle diameter, and distance between the needle tip and grounded collector). Consequently, modifying any or all of these parameters directly affects the size, shape, and morphology of the resulting nanofibers. By exerting control over these parameters, it becomes possible to produce precisely defined fibers tailored for specific applications. Additionally, ensuring control over the structural characteristics of electrospun fibers—such as fiber diameter, porosity, volume ratio, surface morphology, and mechanical properties—is crucial for biomedical applications.

1.2 Significance of fabrication techniques

The fabrication technique used to produce nanofibers plays a crucial role in determining their properties and functionality. The significance of nanofiber fabrication techniques can be summarized as follows:

Control over Morphology: Nanofiber fabrication techniques allow precise control over the morphology of the resulting fibers, including diameter, length, and surface area. This control is essential for tailoring the physical, chemical, and mechanical properties of nanofibers to meet specific application requirements.

High Aspect Ratio: Nanofibers possess an extremely high aspect ratio, with lengths several orders of magnitude greater than their diameters. Fabrication techniques enable the production of nanofibers with uniform diameters and lengths, resulting in a high aspect ratio, which is advantageous for various applications such as filtration, tissue engineering scaffolds, and reinforcement in composite materials.

Large Surface Area: Nanofibers have an inherently large surface area-to-volume ratio due to their small size and high porosity. Fabrication techniques can further enhance the surface area by creating specific morphologies such as porous, hollow, or structured nanofibers. The increased surface area provides more active sites for chemical reactions, adsorption, and interaction with the surrounding environment, making nanofibers suitable for sensing, catalysis, and energy storage applications.

Tunable Properties: Nanofiber fabrication techniques offer the ability to tune the properties of nanofibers by incorporating dopants, additives, or functional materials during the synthesis process. This allows tailoring of the optical, electrical, magnetic, and mechanical properties of nanofibers for specific applications such as optoelectronics, sensors, and energy devices.

Integration and Hybridization: Nanofiber fabrication techniques enable the integration of different materials and the formation of hybrid structures. By incorporating multiple components into nanofibers, it is possible to create multifunctional materials with enhanced properties and synergistic effects. This opens up opportunities for developing advanced devices, such as flexible electronics, wearable sensors, and drug delivery systems.

Scalability and Reproducibility: Fabrication techniques for nanofibers can be scaled up to produce large quantities of fibers with consistent quality and reproducibility. This is crucial for commercial applications where batch-to-batch consistency and production scalability are essential.

Versatility: Nanofiber fabrication techniques are versatile and can be adapted to various materials, including polymers, metals, oxides, and carbon-based materials. This versatility allows the fabrication of nanofibers with diverse compositions and structures, expanding their application potential across different fields.

2. Template-based fabrication techniques

2.1 Electrospinning

2.1.1 Introduction to the electrospinning method

Electrospinning is a simple, proficient and versatile method to produce polymer based nanofibers for numerous applications. In recent times, this method has garnered considerable interest across various domains, despite its invention dating back to 1934 by Anton [1]. Typically, a fundamental electrospinning configuration consists of a high voltage power supply, a syringe needle connected to the power supply, and a counter-electrode collector, as depicted in the schematic diagram shown in **Figure 2**. In the process of electrospinning, a significantly high electrostatic voltage is applied

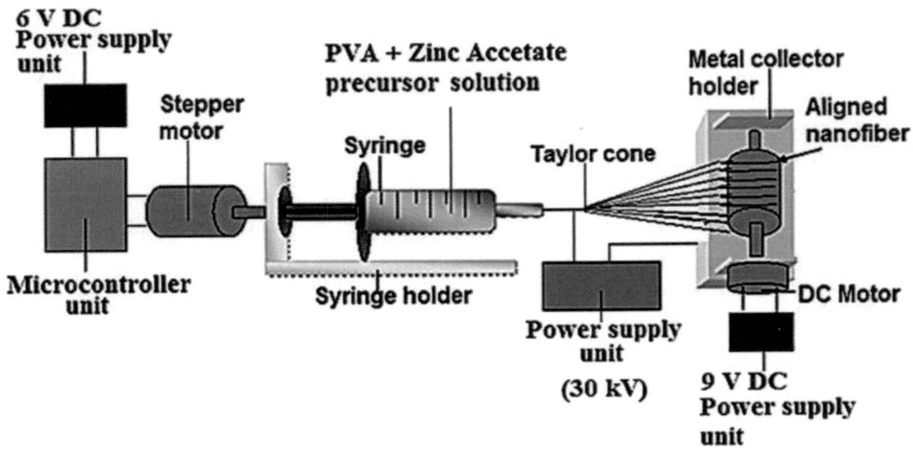

Figure 2.
Schematic diagram of electro-spin coating unit.

to the polymer solution, leading to a substantial electrification of the solution droplet located at the tip of the needle [2, 3]. As a result, the solution droplet at the needle tip receives electric forces, drawing itself towards the opposite collector electrode, thus forming into a conical shape known as "Taylor cone" [4].

When the electric force surpasses the surface tension of the polymer solution, the solution is released from the tip of the "Taylor cone" and forms a polymer jet. This charged jet is then stretched into a fine filament by the strong electric force. By evaporating the solvent within the filament, dry fibers are deposited onto the collector electrode. The electrospinning process and the properties of the fibers are influenced by numerous factors, including the characteristics of the polymer materials (such as polymer structure, molecular weight, and solubility), the properties of the solvent (such as boiling point and dielectric properties), solution properties (such as viscosity, concentration, conductivity, and surface tension), operating conditions (such as applied voltage, collecting distance, and flow rate), and the surrounding environment (such as temperature, gas environment, and humidity). Electrospun nanofibers possess several unique traits, including a high surface-to-mass ratio, excellent interconnectivity of pores within a high porosity structure, flexibility combined with reasonable strength, a wide selection of polymer materials to choose from, the capability to incorporate other materials (such as chemicals, polymers, biomaterials, and nanoparticles) into the nanofibers through electrospinning, and the ability to control secondary structures of the nanofibers to create core/sheath structures, side-by-side structures, hollow nanofibers, and nanofibers with porous structures [5]. These characteristics make electrospun nanofibers suitable for various applications, including filtration, affinity membranes, metal ion recovery, tissue engineering scaffolds, controlled release systems, catalyst and enzyme carriers, sensors, and energy storage [6].

2.2 Software design for nanofiber fabrication

The role of software design is crucial in the development of a software system, as it enables developers to create various models that serve as blueprints for the implementation solution. These models can be thoroughly analyzed and evaluated to determine their ability to meet the specified requirements. Additionally, alternative solutions and trade-offs can be explored and assessed. Ultimately, these models are used to plan subsequent development activities and serve as the foundation for coding and testing processes.

As per the definition provided by the Institute of Electrical and Electronics Engineers (IEEE), design encompasses two aspects: it is both the process of defining the system or component's architecture, components, interfaces, and other characteristics, as well as the outcome of this process. When viewed as a process, software design is the activity within the software development cycle where software requirements are carefully analyzed to generate a description of the internal structure and organization of the system. This description serves as the basis for constructing the system. Specifically, the software design result should outline the system's architecture, including how it is decomposed and organized into components, and it should define the interfaces between these components. Furthermore, it should provide sufficient detail about these components to facilitate their construction.

Software design encompasses two activities that occur between software requirements analysis and software coding and testing. The first activity is software architectural design, also known as top-level design, which involves describing the overall structure and organization of the system and identifying its various components. The second activity is software detailed design, where each component is sufficiently described to enable coding.

The software design process typically follows a series of steps that begin with stating the basic program requirements and gradually adding more detail. This progression typically involves the following stages where it starts with gathering the requirements and specifications. The program's desired operation is initially described at a general level and then in more detail. The next step is gathering the flow of steps, decisions, and loops which is planned, often using diagrams, to indicate dependencies between different subtasks. This stage defines the sequence of operations and the data communication requirements. Next the format of variable types for passing data between functions is determined to design function interfaces. The program's structure and components are designed in a top-down and/or bottom-up design. These two design paradigms help organize the overall structure and individual modules, respectively. Once a plan for the program's appearance has been established, the focus shifts to implementation. The coding process often reveals flaws in the original design or suggests improvements and additional features. Therefore, design and implementation tend to be iterative rather than a linear progression. The final stage involves testing and debugging. Non-trivial programs inevitably contain errors when initially written, so thorough testing is necessary before implementing them in real-time applications. Debugging is performed to identify and rectify any issues.

2.2.1 Mechanism of nanofibers fabrication

The electrospinning technique utilizes a high electric field to generate extremely fine polymeric fibers, ranging in diameter from a few nanometers to a few micrometers. The mechanism behind electrospinning nanofibers is based on the intricate electro-physical interactions between the polymer solution and electrostatic forces. In this process, a high-voltage electric field is established between the injection needle and the collecting screen, facilitated by a power supply and electrodes. As the polymer solution is slowly extruded from the syringe, a semi-spherical droplet of the solution forms at the needle's tip. With the increasing voltage, the charged polymer droplet elongates into a cone shape, accumulating surface charge over time. Once the surface charge surpasses the surface tension of the polymer droplet, a polymer jet is initiated. The solvent within the polymer jet carries the material as it travels towards the collecting screen, further increasing the surface charge on the jet [7]. This rise in surface charge induces instability in the polymer jet as it interacts with the electric field. To counteract this instability, the polymer jet undergoes geometric division, initially forming two jets and subsequently splitting into numerous additional jets as the process continues. The spinning force generated by the electrostatic force on the continuously dividing polymer droplets ultimately leads to the formation of nanofibers. These nanofibers are deposited layer-by-layer on a metal target plate, creating a non-woven nanofibrous mat [8].

Throughout the electrospinning process, both extrinsic and intrinsic parameters are known to impact the structural morphology of the nanofibers [9]. Extrinsic parameters, including environmental humidity, temperature, as well as intrinsic parameters such as applied voltage, working distance, conductivity, and viscosity of the polymer solution, must be optimized to achieve uniform nanofiber production. Within the resulting nanofibrous mat, two main structures are commonly observed: a uniform, continuous fibrous structure or a structure containing beads. The relative abundance of these two structures depends on the contributions of various parameters during the electrospinning process.

2.2.2 Block diagram of nanofibers generator

For the past few decades several researchers have developed electrospinning nanofibers generator with larger unit size and least automation techniques. The design introduced microcontroller for the design of high voltage power supply as well as electrospinning spinneret unit. These new design brought cost reduction, unit portability and automation techniques. This technique has many advantages such as ease of operation and controlling the fiber diameter also functionalizing nanofibers through adding functional agents to the electrospinning solution. Hardware and software designs are required to develop the effective automation.

Figure 3 illustrates a block diagram presenting the configuration of an electrospinning nanofibers generator. This custom-designed generator comprises three main modules: (1) High voltage power supply unit, (2) Spinneret or Plunger unit, and (3) Nanofibers collector unit. The high voltage power supply unit, assisted by a microcontroller, is designed and constructed to provide a variable direct current (DC) source necessary for the production of nanofibers. The spinneret-solution feeding unit is composed of a syringe with a stainless steel needle ranging in size from 0.1 mm to 0.3 mm. It is fitted with a syringe holder and connected to a stepper motor. The stepper motor, controlled by a microcontroller unit, generates precise forward and reverse mechanical forces to move the syringe piston accordingly. This motion facilitates the formation of the Taylor cone, from which aligned nanofibers are fabricated and deposited onto the substrate, securely placed in the metal collector holder. Two types of collectors are employed: flat and drum types.

2.2.3 Design and construction of nanofibers generator modules

The nanofibers generator consists of three modules as shown in **Figure 3**. The first module comprises of DC power supply unit, microcontroller16F877A unit, and stepper motor driver unit. It helps to supply necessary and controlled (Programmed

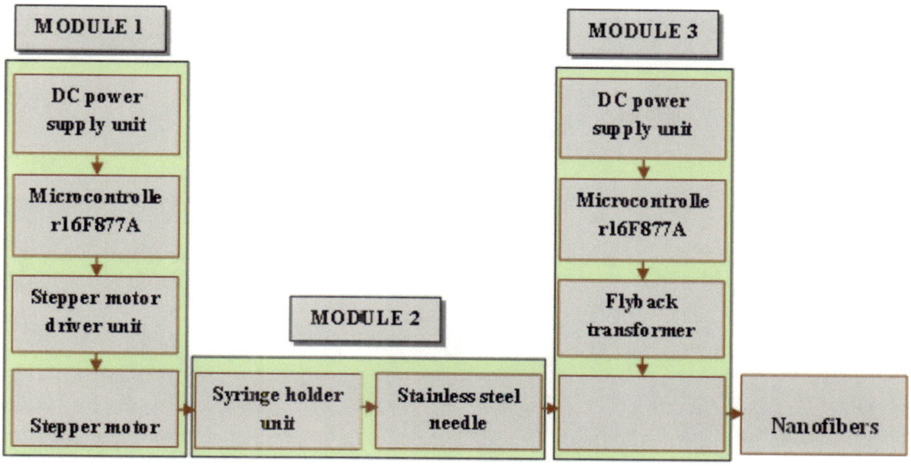

Figure 3.
Block diagram of electrospinning nanofibers generator.

pulses) to the fixed stepper motor and hence the forward and backward movements (in appropriate steps) of the piston connected to the syringe in the next module. The second module consists of syringe holder fitted with syringe and stainless steel needle connected to positive potential of high voltage power supply from module (3). The third module consists of DC power supply unit, microcontroller16F877A unit, stepper motor driver unit, flyback transformer and collector unit which is used to produce 0–30 kV variable DC power. This power is applied across stainless steel needle and collector unit.

2.3 Module I

2.3.1 DC input power source for microcontroller

The first module of the nanofibers generator consists of DC power supply unit. It provides the supply voltage of 5 V DC power to the microcontroller unit. It consists of 0–6 V step down transformer 250 mA, bridge rectifier and voltage regulator (IC-7805). The performance of the unit was checked using Digital Multi Meter (DMM) and Cathode Ray Oscilloscope (CRO). It gives a steady 5 V DC for the load resistance of 5–10,000 Ω. The waveform of generated DC output was studied using storage oscilloscope. The generated AC input waveform and bridge rectifier DC output waveforms are shown in **Figure 4** [10].

2.3.2 Microcontroller 16F877A controlled circuit for spinneret unit

In this research PIC 16F877A is used to move the piston in desired direction and steps to inject the high viscous precursor polymer Sol–Gel for nanofibers fabrication. It is very difficult with high pressure to inject the high viscous polymer through a needle of very small diameter in the order of 0.1–0.3 mm. The unit is designed for the forward and backward movement of piston in this syringe. The microcontroller unit was fitted with three push button switches to stop and control forward and backward movement of the stepper motor. For this requirement, appropriate algorithms were written and the corresponding "C" program is written and compiled to the machine

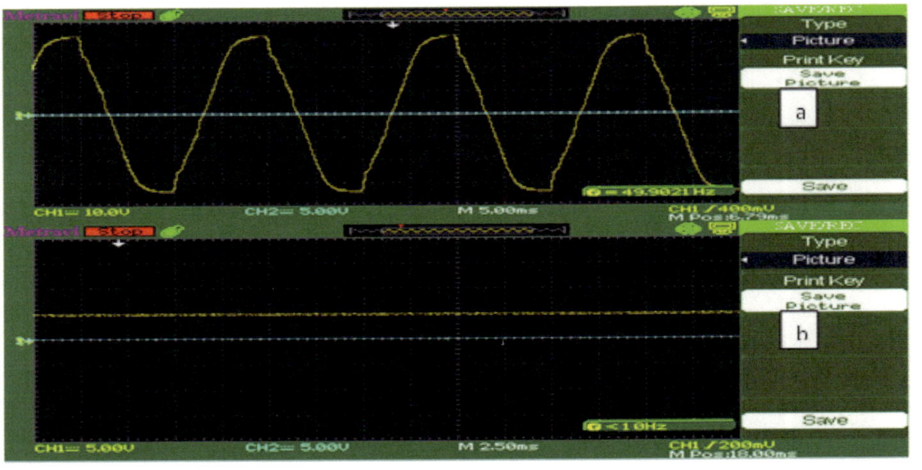

Figure 4.
(a) AC input waveform and (b) DC output waveform of DC power supply.

language and uploaded in the PIC 16F877A using PIC programmer [11]. **Figure 5** depicts the circuit diagram of PIC 16F877A. The developed software program for this operation is attached in the Appendix-B.

2.3.3 Stepper motor driver unit

In order to control the motion of the stepper motor, a driving unit consists of stepper motor driver L 298 is used and the circuit diagram and cooling fan design is as shown in **Figure 6**. It is not possible to operate the stepper motor without the stepper driver circuit. **Figure 7** describes the circuit diagram of stepper motor driver unit.

Necessary voltage for the rpm of stepper motor with high starting pulse can be obtained only using stepper driver unit. For each step movement of the signal is obtained from the microcontroller through this driver unit. L 298 plays a vital role in generating high voltage and high current output for the stepper motor operation.

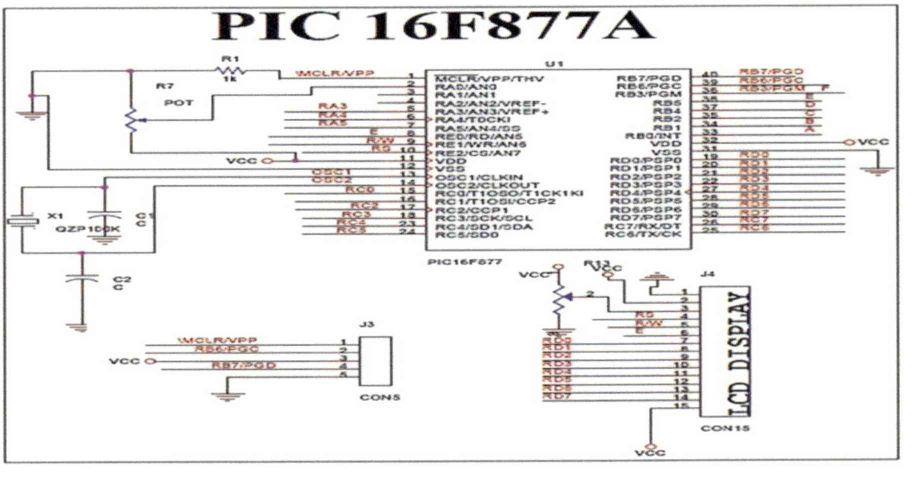

Figure 5.
Circuit diagram of PIC microcontroller 16F877A.

DOI: http://dx.doi.org/10.5772/intechopen.112477

Figure 6.
Circuit diagram of stepper driver photograph.

A micro controller or stepper motor controller can be used to activate the drive transistors in the right order, and this will ease of operation making unipolar motors popular with hobbyists; this is the cheapest way to get precise angular movements.

The L 298 driver circuit can control concurrently up to two stepping motors. Only one stepper motor was used towards this motor driver circuits. It can handle maximum of 4 A, i.e., two amps per motor, however to get the maximum current make sure to add a heat sink on the top of the body [12]. The L 298 has a large cooling brim with a hole in it, making it easy to attach a metal heat sink to it. The motor driver unit system primarily comprises of three significant components namely stepper driver, indexer, and user interface as shown in stepper driver circuit diagram (**Figure 7**).

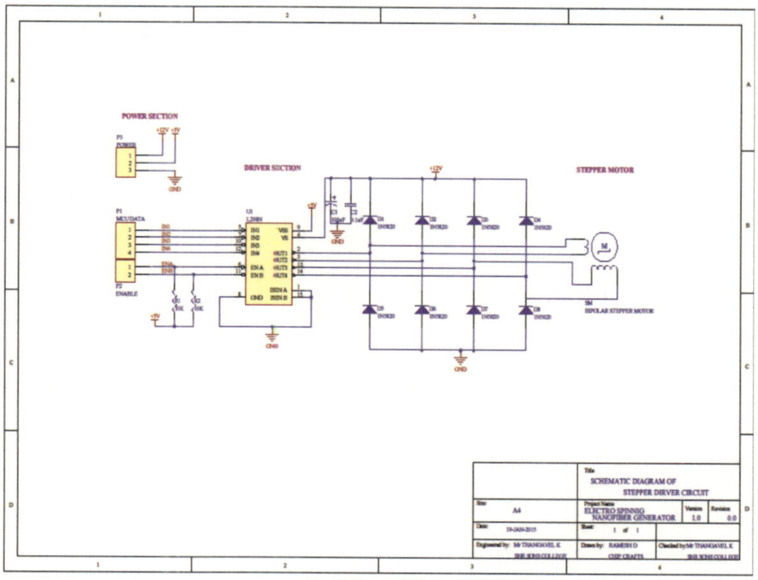

Figure 7.
Stepper driver circuit diagram.

In this research work 23LM C355-20 stepper motor and L 298 stepper motor driver is produced by automation process. The L 298 is a 15—pin multi watt and high power monolithic integrated circuit. This dual full-bridge driver is specifically designed to accommodate TTL logic levels and effectively drive inductive loads like relays and stepping motors, offering both high voltage and high current capabilities. The device includes two enable inputs, which allow for independent enabling or disabling of the driver, regardless of the input signals.

The stepper motor leads were connected in series for a higher inductance and therefore attained higher performance with lower speed. The motor driver will be efficiently able to control and drive a four phase unipolar stepper motor with continuous output current ratings to 1.5 A per phase and 1.5 Ω [13]. The schematic diagram shows a basic connection diagram for controlling motors using the L 298 motor bridge IC. There are three input pins for each motor: Input1, Input2, and Enable1 controls Motor1. Input3, Input4, and Enable2 controls Motor2. **Table 1** describes the input and basic functions of designed stepper driver circuit. In order to activate the motor, the enable1 line must be high. To control the motor and its direction by applying a LOW or HIGH signal to the Input1 and Input2 lines, as shown in **Table 1**.

2.3.4 Stepper motor

Stepper motors are a type of brushless synchronous motors that complete a full rotation cycle in a predetermined number of steps. These motors are specifically designed to be driven by applying excitation pulses to the phase windings, as they cannot be operated simply by connecting the positive and negative leads of a power supply. Instead, they require a stepping sequence generated by a microcontroller to drive their movement. The motor advances in discrete steps according to this sequence. Stepper motors can be classified based on their construction type.

- Variable Reluctance (VR) stepper motor.
- Permanent Magnet (PM) stepper motor.
- Hybrid stepper motor

Enable	Input-I	Input-II	Function
Low	Low	Low	Off
Low	High	Low	Off
Low	Low	High	Off
Low	High	High	Off
High	Low	Low	Motor break
High	High	Low	Motor run forward
High	Low	High	Motor run backward
High	High	High	Motor break

Table 1.
Stepper driver input and function.

Variable reluctance stepper motors are characterized by having a rotor composed of ferromagnetic materials. When the stator is energized, it becomes an electromagnet, exerting a pull on the rotor in that direction. The ferromagnetic material naturally seeks to align itself along the path of minimum reluctance. By energizing the coils, a magnetic field is generated, and the airgap reluctance is adjusted, hence the name "variable reluctance stepper motor." In this type of motor, the motor's direction is independent of the direction of the current flowing in the windings [14].

On the other hand, the rotor of a Permanent Magnet (PM) stepper motor is permanently magnetized. Therefore, the motor's movement is determined by the attraction and repulsion forces between the magnetic poles of the stator and rotor. In this motor, the direction of the motor depends on the direction of the current flowing in the windings since the magnetic poles are reversed by changing the current flow through the rotor. The hybrid stepper motor, as the name suggests, combines the advantages of both the permanent magnet stepper motor and the variable reluctance stepper motor to provide improved efficiency [15]. These two types of stepper motors are the most commonly used. The main distinction lies in the fact that the rotor of the variable reluctance stepper motor is made of a ferromagnetic material, whereas the rotor of the permanent magnet stepper motor is permanently magnetized. The full view and cut view of the original stepper motor are depicted in **Figure 8a** and **b**.

2.4 Module II

2.4.1 Syringe holder unit or spinneret unit

Syringe pumps, also known as spinnerets, are small pumps precisely operated by stepper motors. Microcontroller 16F877A drives the stepper motor using stepper driver to generate necessary mechanical force [16]. The generated mechanical force is used to move the syringe tap on either direction. During forward motion, the mechanical force moves the plunger of the syringe in the forward direction resulting in pushing the Sol–Gel to the syringe needle tip, where as in the reverse direction the mechanical force generated by the stepper motor brings the plunger of the syringe to

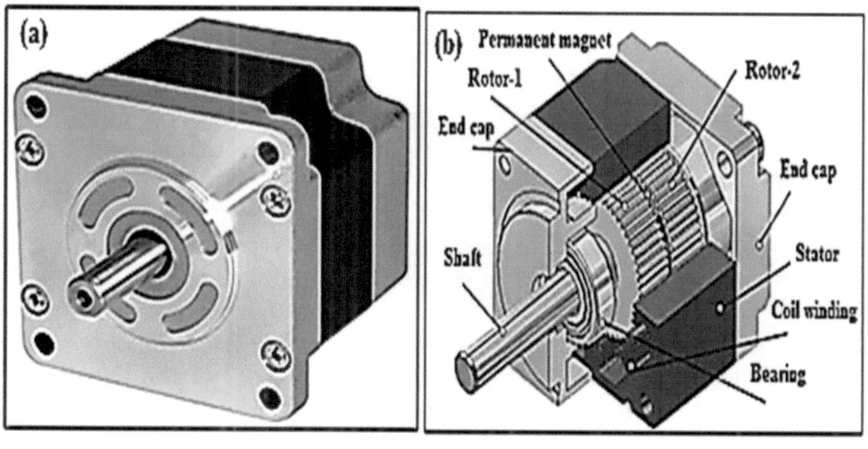

Figure 8.
(a) Stepper motor full view and (b) cut view.

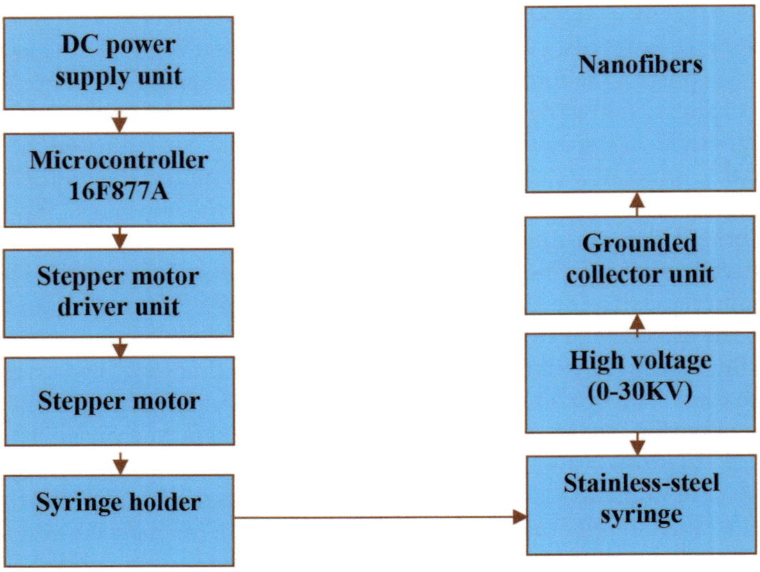

Figure 9.
Block diagram of a syringe holder unit/spinneret unit.

its initial position for refilling. The microcontroller16F877A is programmed using "C" language to generate a controlled forward and reverse mechanical force. This software enabled microcontroller accurately generates controlled mechanical forces in forward duration and reverse duration. By varying software code in the microcontroller the mechanical force duration can be varied. A screw rod is connected in between the stepper motor and syringe needle, due to this set up, the system requires a high power DC supply. The 5 Amps 12 Volt DC source is used to energize the stepper motor via stepper driver circuit. Cooling fans are installed to dissipate the enormous amount of heat generated during the generation of mechanical force to move the plunger of the syringe and to keep the driver unit at optimal temperature. **Figure 9** describes the block diagram of spinneret unit.

3. Conclusion

- Hardware and software design of DC power supply unit, microcontroller unit, MOSFET switching circuit, high voltage power supply unit and spinneret unit design has been completed.

- Electronic components and circuit boards needed in the proposed research have been systematically identified and incorporated in this research.

- Microcontroller based high voltage power supply and syringe holder unit has been designed by using PIC16F877A. A variable frequency from 250 Hz to 25 kHz can be generated for the high voltage. Inbuilt PWM generator has been used to generate pulse of high voltage with a variable frequency.

- The "C" language software works as expected. High DC voltage output can be configured from PIC16F877A interface software via serial communication.

The observed output voltages were varied from 0–30 kV with the help of variable potentiometer and its accurate results were displayed on liquid crystal display screen (LCD).

- Two different metal collectors were initially designed and studied, both collectors yield best results. It provides uniform fibers on the metal collector plate with good morphological characteristics. It is also observed that there is no bead like structure found in the fiber obtained with both type of collectors.

- Polymer flow rate can be configured from microcontroller interface software via control circuits in the main board.

- The overall results show that voltage, frequency and duty cycle are controlled by the resistance of the switching circuit, and the output voltage proportionally changes by changing the resistance.

- The electrospinning generator design has been optimized and formulated.

- This improvised microcontroller controlled nanofibers generator has a number of advantages over conventional techniques: Simple, cost effective, portable and able to fabricate continuous nanofibers with small scale production.

Author details

Thangavel Krishnasamy*, Balaprakash Vadivel, Prema Rangasamy and Mahitha Mohan
Department of Electronics, Hindusthan College of Arts & Science, Coimbatore, Tamil Nadu, India

*Address all correspondence to: krishnavel1970@gmail.com

References

[1] Huang ZM, Zhang YZ, Kotaki M, Ramakrishna S. A review on polymer nanofibers by electrospinning and their applications in nanocomposites. Composite Science and Technology. 2003;**63**(15):2223-2253

[2] Ahn YC, Park SK, Kim GT, Hwang YJ, Lee CG, Shin HS, et al. Development of high efficiency nanofilters made of nanofiber. Current Applied Physics. 2006;**6**(6):1030-1035

[3] Baumgarten PK. Electrostatic spinning of acrylic microfibers. Journal of Colloid and Interface Science. 1971;**36**(1):71-79

[4] Doshi A, Reneker DH. Electrospinning process and applications of electrospun fibers. Journal of Electrostatics. 1995;**35**(2-3):151-160

[5] Larrondo L, Manley RSJ. Electrostatic fiber spinning from polymer melts, experimental-observations on fiber formation and properties. Journal of Polymer Science: Part A. 1981;**19**(6):909-920

[6] Xin Y, Huang Z, Li W, Jiang Z, Tong Y, Wang C. Core-sheath functional polymer nanofibers prepared by co-electrospinning. European Polymer Journal. 2008;**44**(4):1040-1045

[7] Deitzel JM, Kleinmeyer J, Harris D, Beck Tan NC. The effect of processing variables on the morphology of electrospun nanofibers and textiles. Polymer. 2001;**42**(1):261-272

[8] Thoppey NM, Gorga RE, Clarke LI, Bochinski JR. Control of the electric field e-polymer solution interaction by utilizing ultra-conductive fluids. Polymer. 2014;**55**:6390-6398

[9] Reneker DH, Chun I. Nanometer diameter fibers of polymer, produced by electro-spinning. Nanotechnology. 1996;**7**(3):216-223

[10] Feng L, Li S, Li H. Super-hydrophobic surface of aligned polyacrylonitrile nanofibers. Angewandte Chemie International Edition. 2002;**41**(7):1221-1223

[11] Lee KH, Kim HY, Bang HJ, Jung YH, Lee SG. The change of bead morphology formed on electrospun polystyrene fibers. Polymer. 2003;**44**(14):4029-4034

[12] Zhang C, Yuan X, Wu L, Han Y, Sheng J. Study on morphology of electrospun poly (vinyl alcohol) mats. European Polymer Journal. 2005;**41**(3):423-432

[13] Reneker DH, Chun I. Nanometer diameter fibers of polymer, produced by electrospinning. Nanotechnology. 1996;**7**(3):216-223

[14] Buchko CJ, Chen LC, Shen Y, Martin DC. Processing and microstructural characterization of porous biocompatible protein polymer thin films. Polymer. 1999;**40**(26):7397-7407

[15] Demir MM, Yilgor I, Yilgor E, Erman B. Electrospinning of polyurethane fibers. Polymer. 2002;**43**(11):3303-3309

[16] Yordem OS, Papila M, Menceloglu YZ. Effects of electrospinning parameters on polyacrylonitrile nanofibers diameter: An investigation by response surface methodology. Materials and Design. 2008;**29**(1):34-44

Chapter 5

Fabrication of Variable Morphologies on Argon Sputtered PMMA Surfaces

Divya Gupta, Rimpi Kumari, Amena Salim, Rahul Singhal and Sanjeev Aggarwal

Abstract

Ion beam induced patterning and fabrication of various topographies over polymeric surfaces has drawn strong interest due to latent applications in photonics, magnetic devices, optical devices and photovoltaics etc. In this work, we report the controlled surface structuring and evolution of different morphologies in Poly(methyl methacrylate) polymer using Ar^+ ion beam fabrication technique. Morphological and structural analysis has been performed by ex situ Atomic Force Microscopy (AFM) and X-ray Diffraction. The effect of oblique incidences on argon sputtered films was evaluated by various surface topography and texture parameters, such as Fast Fourier Transforms, surface roughness, skewness, kurtosis. AFM study demonstrates fabrication of transient morphologies over argon sputtered surfaces. One dimensional (1D) cross section scans of surface profiles are determined and morphological features are investigated. The results showed halo peaks in the XRD patterns, which indicate the amorphous nature of this type of polymer. The formation of these surface structures is attributed to the different degree of sputtering yield at different off-normal incidences and preferential sputtering of hydrogen in comparison to carbon in ion sputtered surfaces.

Keywords: PMMA, argon ion irradiation, atomic force microscopy, X-ray diffraction, sputtering

1. Introduction

Surface structuring and patterning of polymers by ion irradiation is a technique that involves exposing a polymer surface to ion beam to selectively modify its structural and chemical properties. This technique allows the creation of complex patterns with high resolution and precision, making it highly valuable in a range of fields, such as microelectronics, sensors, and biomedical applications [1–10].

During the process of ion irradiation, the incident ions penetrate the polymer surface and cause a series of physical and chemical changes, such as chain scission, cross-linking, and the formation of new chemical groups. These changes can be controlled by adjusting the beam parameters such as ion energy, ion dose, and angle of incidence, allowing the creation of a range of patterns with different geometries and varying depths.

One of the key advantages of ion irradiation induced structuring and patterning is its ability to create highly uniform patterns over a large area, making it ideal for large-scale manufacturing. Furthermore, this technique can be applied to a range of polymers, including those that are difficult to pattern using conventional lithography techniques [1–11].

Ion beam induced patterning and structuring in materials is a rapidly evolving field with a wide range of applications, including micro- and nanofabrication, surface modification, and materials engineering. Reports are available in the existing literature describing the use of this versatile technique for creation of diverse morphologies over different classes of materials [12–23].

Vazquez et al. [12] have given an up-dated account of the progress reached when surface composition plays a relevant role, with a main focus on ion beam irradiation surface patterning with simultaneous co-deposition of foreign atoms. Further, they have reviewed the advances in ion beam irradiation of compound surfaces as well as ion beam irradiation systems where the ion employed is not a noble gas species. Additionally, they have explained the main theoretical models aimed at describing these nanopattern formation processes. Lastly, they have addressed two main special features of the patterns induced by this ion beam irradiation technique, namely the enhanced pattern ordering and the possibility to produce both morphological and chemical patterns.

Cuerno et al. [13] have presented a perspective investigation on the main developments that have led to the current understanding of nanoscale pattern formation at surfaces by ion-beam irradiation, from the points of view of experiments, applications, and theory, and offer an outlook on future steps that may eventually facilitate full harnessing of such a versatile avenue to materials nanostructuring.

They have started with a brief survey on the key issues which have been dealt with up to the recent past with respect to both experiments and applications and theoretical aspects of IBS surface nanopatterning. Then, they have assessed the current status of open problems which we consider important, again providing perspectives on these from the points of view of experiments and applications and also from the theoretical and computational viewpoints. For definiteness, they have restricted the discussion to the patterns formed by low energy ($E_{ion} \lesssim 10$ keV), broad ion beams, albeit with occasional mention of results at higher energies.

Valbusa et al. [14] have studied that the surface etching by ion sputtering can be used to pattern surfaces. Recent studies using the high-spatial-resolution capability of the scanning tunneling microscope revealed in fact that ion bombardment produces repetitive structures at nanometer scale, creating peculiar surface morphologies ranging from self-affine patterns to fingerprint-like and even regular structures, for instance waves (ripples), chequerboards or pyramids. The phenomenon is related to the interplay between ion erosion and diffusion of adatoms (vacancies), which induces surface re-organization. Their paper reviews the use of sputter etching to modify 'in situ' surfaces and thin films, producing substrates with well-defined vertical roughness, lateral periodicity and controlled step size and orientation.

Li et al. [15] have studied the recent Applications of Ion Beam Techniques on Nanomaterial Surface Modification and hence design of Nanostructures and Energy Harvesting. According to them, ion beam techniques have extensively been applied for modulating the performance of various nanomaterials. In addition, ion beam techniques have also been used to fabricate nanomaterials, including 2D materials, nanoparticles, and nanowires. In addition, ion beam techniques exhibit high controllability and repeatability. The recent progress in ion beam techniques for nanomaterial surface modification is systematically summarized and existing challenges and potential solutions are presented.

Gupta et al. [16] have studied the controlled surface modification and nanodots structures over Si(111) surfaces produced by oblique angle sputter deposition of 80 keV Ar + beam. Temporal parameters such as self-assemble, tunability of size and density of fabricated nano-dots exhibited distinct fluence dependence. Crystalline to amorphous (c/a) phase transition for sputter deposited Si(111) surfaces has been observed. RBS/C reveals the non-linear response of damage distribution with argon ion fluence. The underlying self-organization mechanism relied on ion beam sputtering induced erosion and re-deposition of Si atoms thereby leading to mass transport inside the created amorphous layers.

They have [17] also studied the controlled surface structuring of amorphous silicon carbide (a-SiC) thin films as grown on silicon substrates Si(111) by Radio Frequency (RF) sputtering and irradiation of obliquely incident 80 keV Ar^+ ion beam has been investigated. Sub-wavelength ripple patterns with wave-vector parallel to the ion beam projection are found to evolve on argon sputtered surfaces. Studies reveal that the temporal parameters such as ripple wavelength and amplitude, ordering and homogeneity of these patterns vary non-linearly with argon ion fluence. The formation of such surface structures is attributed to the preferential sputtering of silicon in comparison to carbon.

Goyal et al. [19, 20] have studied the temporal variations in nano-scale surface morphology generated on Polypropylene (PP) substrates by 40 keV argon ion sputtering. Formation of ripple patterns and its transition to dot morphology has been realized. Switching of smoothing mechanism from ballistic drift to ion enhanced surface diffusion is the most probable cause for such morphological transition.

Kumari et al. [21] have investigated the patterning of HDPE surfaces by oblique argon ion irradiation. Formation of ripple patterns and its transition to sideways bars & valleys has been observed. Further, transition from parallel to perpendicular mode patterns has been seen. Curvature dependent sputtering and role of various smoothing processes describes the observed surface patterning.

Overall, these studies demonstrate the diverse and exciting potential applications of ion beam-induced patterning and structuring in materials, and highlight the ongoing development of new techniques and mechanisms to control and manipulate these structures.

Hence, motivated by the universality of this IBS technique and pivotal applications of patterned polymeric surfaces, here, we have made an attempt in present endeavor to explore the controlled patterning of poly(methyl methacrylate) PMMA surface by obliquely incident Ar^+ beam erosion with optimized ion beam parameters.

2. Materials and methods

The PMMA studied, was procured from Goodfellow Ltd. (UK). The samples used were flat rectangular wafers of 1 mm thickness.

These specimens were then sputtered with 30 keV Ar^+ ions at an oblique incidence of 0°, 15°, 30°, 40°, 50°, 75° with respect to sample surface. The ion beam sputtering experiments were carried out at room temperature using 200 kV Ion Accelerator [24], which is available in Ion Beam Centre, KUK. The beam current density of 0.5 $\mu A/cm^2$

was used to sputter the samples. The argon fluence was kept constant at 2×10^{16} ions/cm^2.

The morphological evolution of these sputtered surfaces has been examined by Atomic Force Microscopy (AFM) utilizing Bruker HR Multimode-8 instrument. The surfaces were scanned in scan-asyst tapping mode with Antimony(n) doped Si cantilever tip. Different types of scans were obtained at random places on the sample surface. The scanned areas (10×10 μm^2) were sampled at 256 equidistant points at low scan rate of 1 Hz. The scanned images were examined using NanoScope 1.8 software. The observed surface features were evaluated using section analysis and Fast Fourier Transformations (FFTs) of corresponding AFM images.

Structural analysis has been performed employing Bruker AXS D8 Advance X-ray diffractometer using Cu Kα X-ray (λ = 1.5406 Å).

3. Results and discussion

3.1 Morphological analysis

Figure 1a–g presents the two dimensional (2D) topography image of the un-irradiated and 30 keV Ar^+ sputtered PMMA surface at various incident angles of 75° to 15° keeping ion fluence fixed at 2×10^{16} Ar^+ cm^{-2}. The scanned size for these images is 10×10 μm^2. The consequent FFT (Fast Fourier Transform) of micrographs has also been illustrated in 2D AFM images. The qualitative analysis of these AFM images in terms of temporal parameters (obtained through section analysis of respective AFM images) have been calculated using Nanoscope 1.8 software and are presented in **Table 1**.

It is clear from **Figure 1a** that the surface is smooth in nature. Corresponding Fast Fourier Transform (FFT) reveals the smoothness of the un-treated PMMA polymer surface.

Prominent changes in surface morphology can be easily trailed from the comparison of AFM micrograph of **Figure 1a** with **Figure 1b–g**.

After argon ion irradiation at an off-normal incidence of 90°, the rough surface transform into pits with irregular hills. The corresponding FFT image depicts the increase in ordering and homogeneity of the evolved surface features.

With a subsequent decrease in angle of incidence to 75°, the hill morphology becomes prominent with simultaneous occurrence of pits (**Figure 1c**). The enhanced ordering of the evolved hill morphology can be easily deduced from the respective FFT image as shown in inset.

Fascinatingly, with further decrease in angle of incidence to 50°, irregular wavy features in place of pits on the surface can be easily seen in **Figure 1d**. Also, change in the preferred orientation of the surface features can be inferred from the inset image.

As the sputtering proceeds and angle of incidence decreases to 40°, the ripple morphology becomes prominent and pit morphology disappears. These wavy features have orientation parallel to the ion beam direction. The increase in ordering of ripple morphology can be easily observed from the inset images.

With prolonged decrease in angle of incidence to 30°, similar behavior has been observed. Here, also early stages of ripple formation can be clearly seen from the AFM image as shown in **Figure 1f**. This ripple morphology has wave-vectorial orientation parallel to the ion beam direction. Enhanced ordering among the evolved surface structures can be easily deduced from the respective FFT image.

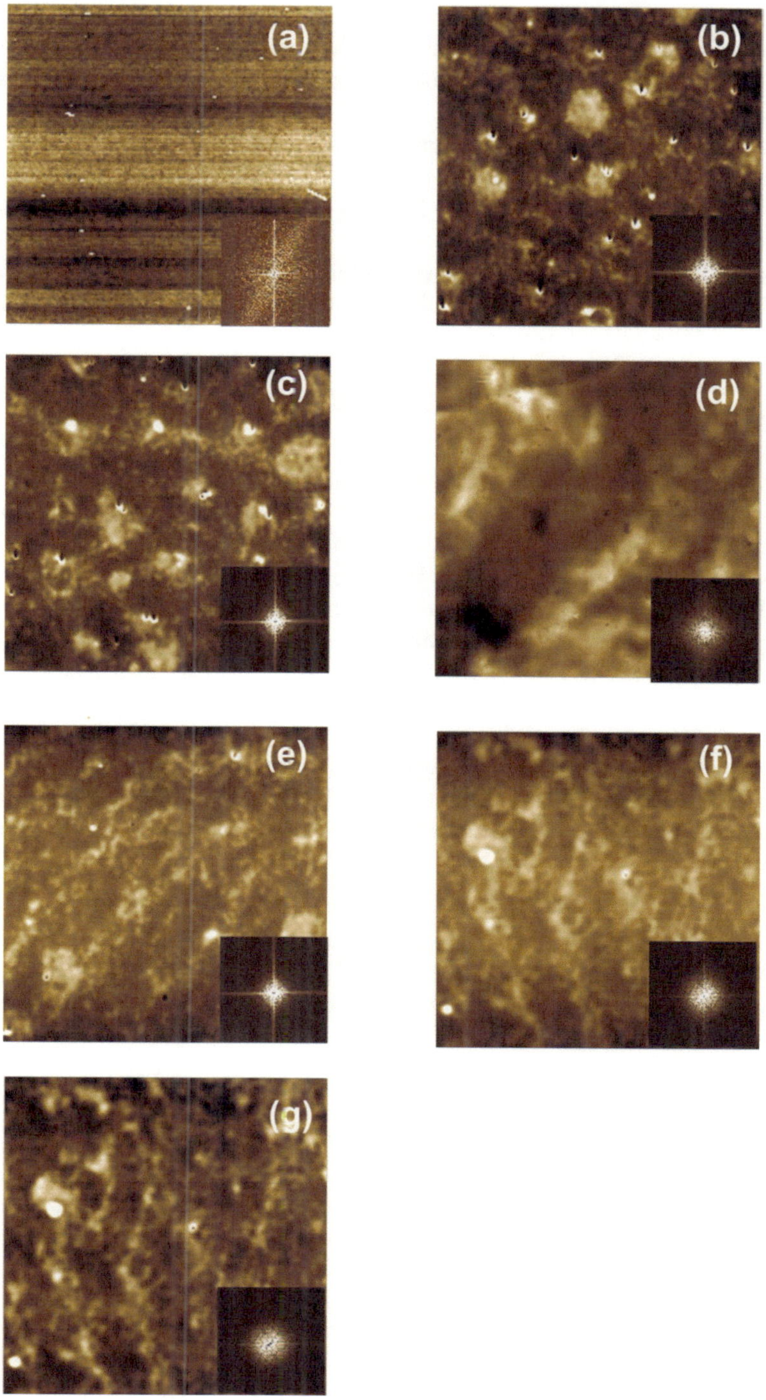

Figure 1.
2D AFM micrographs of for (a) untreated PMMA and irradiated with 30 keV argon ions to a fluence of 2×10^{16} ions/cm^2 at oblique incidences of (b) 90°, (c) 75°, (d) 60°, (e) 50°, (f) 40°, (g) 15° with respect to sample surface.

Similar behavior of surface morphology can be observed at lowest oblique incidence of 15°. Presence of wavy patterns over the sample surface can be easily seen from the AFM image as shown in **Figure 1g**. corresponding FFT image reveals the ordering of the surface features.

Figure 2 presents the one-dimensional line scans of AFM micrographs for PMMA irradiated with 30 keV argon ions to a fluence of 2×10^{16} ions/cm^2 at oblique incidences of (a) 90°, (b) 75°, (c) 60°, (d) 50°, (e) 40°, (f) 15° with respect to sample surface.

It is clear from **Figure 2** that for normal incidence of argon ions, the surface morphology is majorly dominated by pits. Thereafter, gradual switching to wavelike morphology can be easily inferred from the line scans.

3.2 Qualitative description

Surface RMS roughness is a common parameter used to quantify the surface roughness of a sample in Atomic Force Microscopy (AFM) images. It is a statistical measure that describes the average deviation of the height of the sample surface from its mean value over a given area [1–7]. Hence, keeping this in mind, we have calculated the RMS roughness by Nanoscope 1.8 software. The surface roughness for these sputtered surfaces is summarized in **Table 1**.

Moreover, skewness is a statistical measure that describes the degree of asymmetry in a distribution of data. In general, a positive skewness value indicates that the height distribution is skewed to the right, meaning that there are more high points on the surface than low points. A negative skewness value indicates that the height distribution is skewed to the left, meaning that there are more low points on the surface than high points. A skewness value of zero indicates that the height distribution is

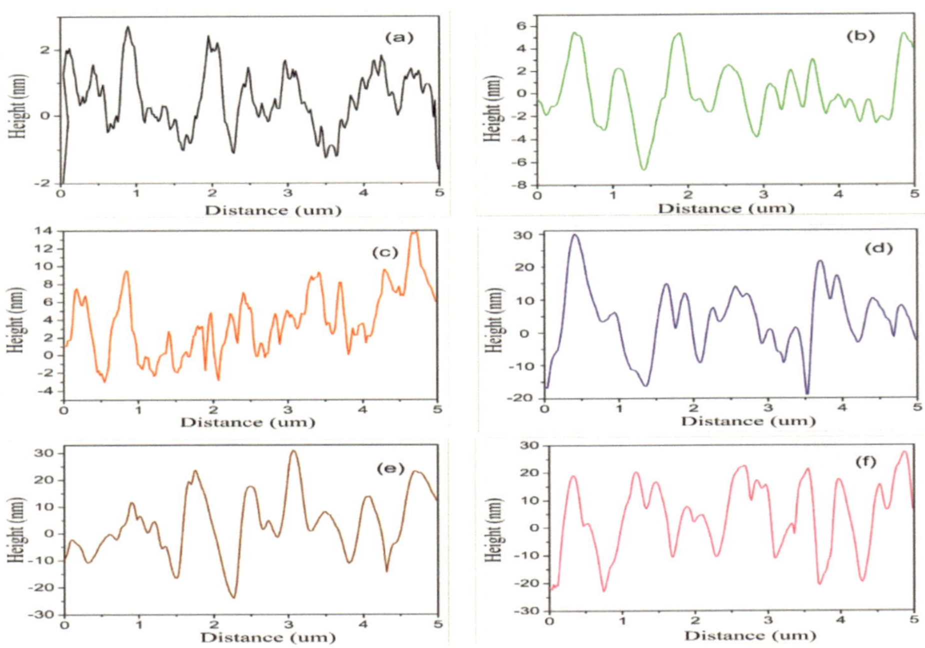

Figure 2.
Time evolution of the line scans with the ion incident angle.

S. No.	PMMA sputtered with oblique incidence of	RMS Roughness (nm)	Skewness	Kurtosis
1.	Untreated	03 ± 0.20	0.11	3.04
2.	90°	10.5 ± 1.20	0.23	3.25
3.	75°	13.1 ± 1.40	0.25	3.18
4.	60°	24.32 ± 3.20	0.26	3.08
5.	50°	15.8 ± 2.30	0.42	3.36
6.	40°	12.61 ± 1.80	0.44	3.34
7.	15°	11 ± 1.60	0.48	3.38

Table 1.
Values of RMS roughness, skewness and kurtosis as a function of angle of incidence.

symmetric, meaning that there are equal numbers of high and low points on the surface [1–5].

Kurtosis is a statistical measure that characterizes the shape, or "peakedness," of a distribution. In the context of Atomic Force Microscopy (AFM) images, kurtosis can be used to analyze the surface roughness and topography of the sample being imaged [2–6].

Specifically, the kurtosis of an AFM image quantifies the degree to which the height distribution of the image deviates from a normal or Gaussian distribution. A high kurtosis value indicates that the height distribution has a sharp peak and heavy tails, while a low kurtosis value indicates a flatter distribution with less pronounced tails [1–7].

By analyzing skewness, kurtosis along with RMS roughness, a more complete picture of the surface morphology has been obtained and these parameters as a function of angle of incidence are listed in **Table 1.**

Table 1 clearly reveals that the surface roughness varies non-linearly with the angle of incidence of argon ions. The surface roughness for untreated PMMA surface is found to be 03 ± 0.20 nm which again confirms the smoothness over the sample surface. Interestingly, with sputtering at normal incidence, the surface roughness increases abruptly. This in turn reveals the roughening of the initial smooth surface. RMS roughness increases continuously with decrease in ion incidence to 60°. This clearly point towards the roughening of the PMMA sample surface. Thereafter, the surface roughness decreases leading to smoothening of the surface.

Moreover, positive value of skewness reveals the presence of hill morphology or ripple morphology above the sample surface. Furthermore, value of nearly 3 for kurtosis indicates the peakedness of the observed morphology over the sample surface.

3.3 Structural analysis

The X-ray diffraction patterns of untreated and 30 keV argon ion irradiated PMMA specimens with a fluence of 2×10^{16} ions/cm^2 at oblique incidences of 90°, 75°, 60°, 50°, 40°, 15° with respect to sample surface are shown in **Figure 3**.

The diffraction pattern of un-treated PMMA specimen shows a broad and intense peak centered at 14.24° with one lower intensity signal centered at 29.70°. The presence of later signal is attributed to the diffuse scattering of amorphous PMMA specimen.

The shape of the first main signal indicates the ordered packing of the polymer chains. The intensity and shape of the second signal are attributed to the effect

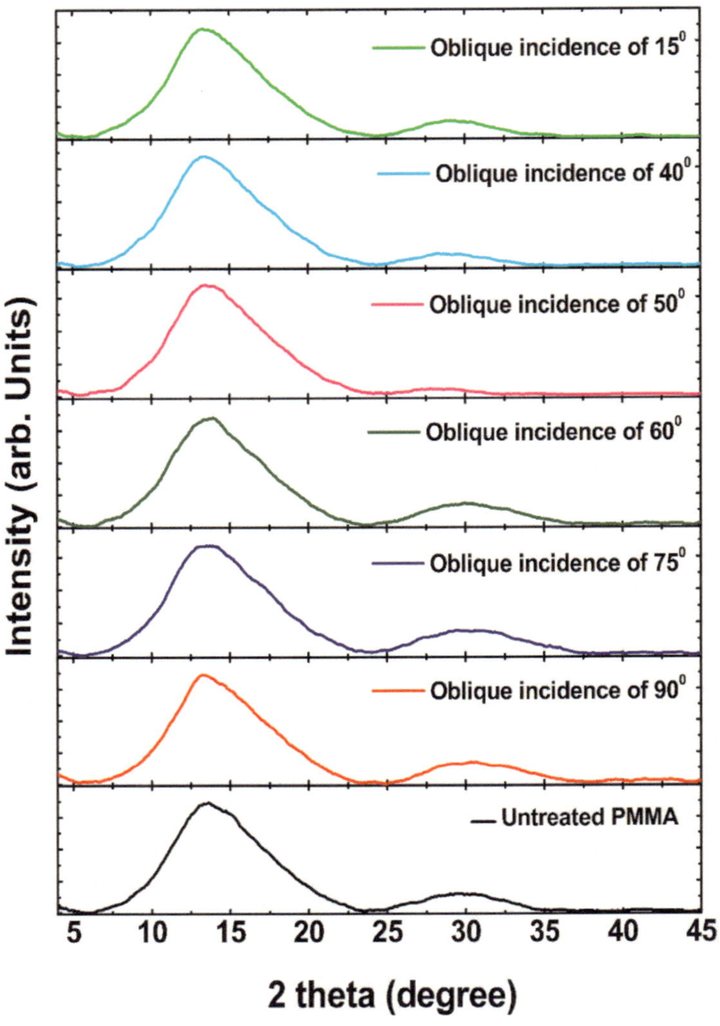

Figure 3.
X-ray diffraction pattern for untreated PMMA and irradiated with 30 keV argon ions to a fluence of 2×10^{16} *ions/cm*^{2} *at oblique incidences of 90°, 75°, 60°, 50°, 40°, 15° with respect to sample surface.*

occurring inside the main chains in the polymeric matrix. The observed broad humps in the XRD spectrum indicate the presence of crystallites of very low dimensions. Further, the absence of prominent peaks in this XRD analysis indicates the amorphous nature of the PMMA polymer [25, 26].

The XRD pattern for PMMA samples sputtered with 30 keV argon ions shows no appreciable change in the peak position and Intensity.

For complete insights into the structural behavior, the crystallite size (a) has been determined using Scherrer's formula [25].

$$a = \frac{0.94\lambda}{\beta cos\theta} \tag{1}$$

Where β is the full width at half maxima (FWHM), λ is the wavelength of the x-ray beam and θ is the diffraction angle.

Values of crystallite size have been summarized in **Table 2**.

Table 2 reveals that the crystallite size varies non-linearly with angle of incidence of argon ions. Crystallite size is always smaller pointing towards the amorphous nature of the surface region.

3.4 TRIM simulations

TRIM (Transport and Range of Ions in Matter), simulation code, is widely used to study the effect of energetic ion bombardment on the different properties of materials, such as surface erosion, sputtering, and implantation [27].

In the present work, we have performed these TRIM simulations to calculate the projected range of incident argon ions at different ion incidences with respect to surface normal and sputtering yield of different components of target matrix.

Figure 4 presents the variation in energy transferred to recoiling atoms for different ion incidences of argon.

It is clear from **Figure 4** that with decrease in angle of incidence, the energy transferred to recoiling atoms increases. This clearly point towards more surface disordering with decrease in angle of incidence.

It can be inferred from **Table 3** that for normal incidence, the incoming argon ions penetrate deep into the polymeric matrix. The projected range is maximum for normal incidence of argon ions. As the angle of incidence of incoming ions decreases, the projected range decreases. For lowest oblique incidence, projected range is found to be the lowest. Interestingly, the sputtering yield increases with decrease in angle of incidence of the argon ions. Sputtering yield is found to be maximum for lowest oblique incidence under present case.

The observed morphological and structural behavior can be explained in terms of sputtering yield, energy to recoils and projected range variations with angle of incidence. Basically, two phenomena occur upon energetic ion irradiation of polymeric matrix: one is the scissioning of polymeric chains which results in bond breaking and shortening of chains predominantly occur as a result of nuclear energy loss process and the other is cross-linking and conjugation mainly due to electronic energy loss process [1]. Both these processes correspond to the roughening and smoothing of polymeric surface which consequences in spontaneous pattern formation triggered by surface instability.

Angle of incidence	2θ (°)	FWHM (°)	a (nm)
Untreated	14.24	7.13	1.17 ± 0.01
90°	14.25	7.07	1.18 ± 0.02
75°	14.24	7.38	1.13 ± 0.01
60°	14.23	7.30	1.15 ± 0.02
50°	14.14	7.20	1.16 ± 0.01
40°	14.13	7.16	1.17 ± 0.02
15°	14.13	7.10	1.18 ± 0.03

Table 2.
Values of Bragg angle, FWHM and crystallite size as a function of angle of incidence.

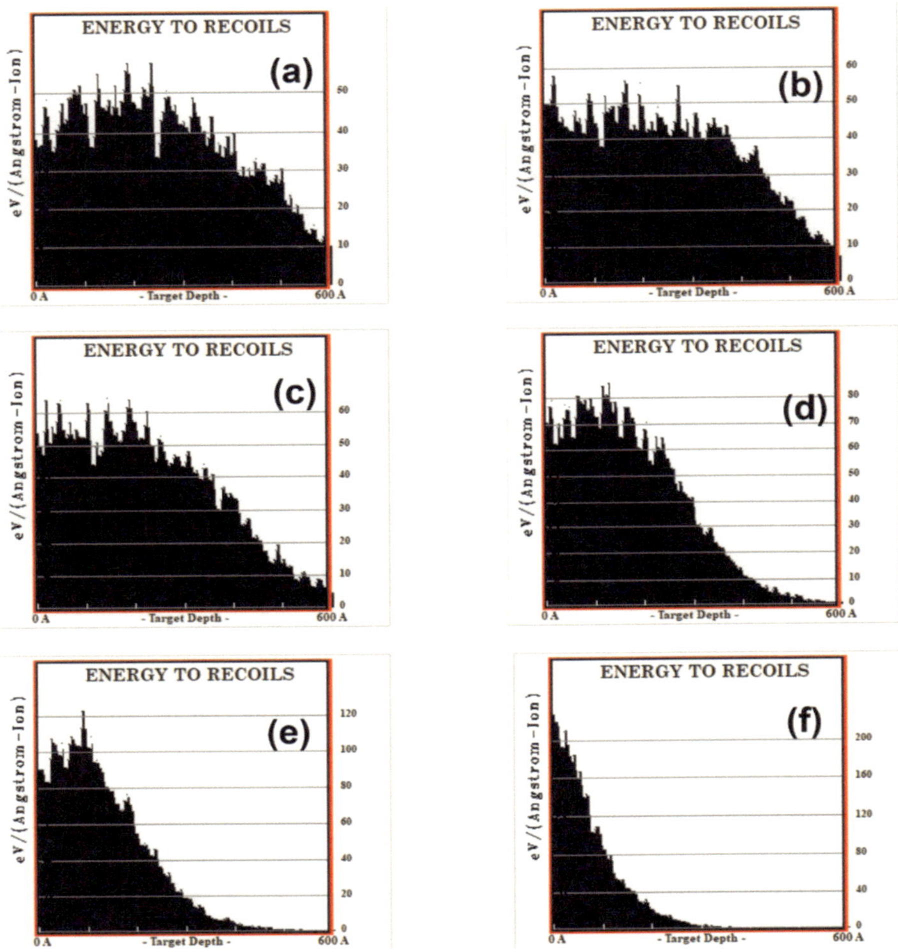

Figure 4.
Energy transferred to recoiling atoms for different argon ion incidences.

PMMA sputtered at ion incidence of	Projected range (nm)	Sputtering yield of hydrogen (atom/ion)	Sputtering yield of carbon (atom/ion)	Sputtering yield of oxygen (atom/ion)
90°	51.4 ± 14.3	1.06	0.22	0.24
75°	49.6 ± 14.2	1.18	0.26	0.27
60°	44.4 ± 13.6	1.80	0.46	0.45
50°	39.3 ± 13.1	2.55	0.69	0.64
40°	33.0 ± 12.4	3.86	1.11	0.96
15°	16.2 ± 08.8	14.34	4.58	3.74

Table 3.
Values of projected range, sputtering yield of hydrogen, carbon and oxygen as a function of angle of incidence.

For normally incident argon ions, the projected range is more and energy transferred to recoiling atoms is less. This results in more structural damage of the polymeric matrix leading to the formation of pit morphology. Interestingly, lower sputtering yields of carbon, hydrogen and oxygen further adds to the formation of pits inside the surface due to more disordering induced inside the surface.

With decrease in angle of incidence to 75°, the projected range decreases and energy to recoils increases. This further produces pit morphology but with less pronounced effect. The increase in sputtering yield of carbon, hydrogen and oxygen causes more surface disordering leading to the formation of pit morphology but majorly in the surface region.

With further decrease in oblique incidence, the projected range deceases gradually and energy transferred to recoiling atoms increases significantly. The increase in sputtering yield of carbon, hydrogen and oxygen has been observed. All these factors result in more surface disordering leading to wave like patterns in place of pit morphology.

At lowest oblique incidence, projected range decreases while energy transferred to recoils increases. Further, sputtering yield increases abruptly leading to only alterations in surface region. Hence, prominent wavy patterns are observed at this stage under present experimental conditions.

4. Conclusions

In summary, we report the controlled surface structuring and evolution of different morphologies in Poly(methyl methacrylate) (PMMA) polymer using Ar^+ ion beam fabrication technique. Morphological and structural analysis has been performed by ex situ Atomic Force Microscopy (AFM) and X-ray Diffraction. The effect of oblique incidences on argon sputtered films was evaluated by various surface topography and texture parameters, such as Fast Fourier Transforms, surface roughness, skewness, kurtosis. AFM study demonstrates fabrication of transient morphologies over argon sputtered surfaces. One dimensional (1D) cross section scans of surface profiles are determined and morphological features are investigated. The results showed halo peaks in the XRD patterns, which indicate the amorphous nature of this type of polymer. The formation of these surface structures is attributed to the different degree of sputtering yield at different off-normal incidences and preferential sputtering of hydrogen in comparison to carbon in ion sputtered surfaces.

Acknowledgements

This research was primarily supported by the Department of Science and Technology (DST), New Delhi for funding major research project for utilizing 200 kV Ion Accelerator and related characterization facilities. Authors are thankful to the Ministry of Human Resource and Development (MHRD) for RUSA 2.0 grants for Centre for Advanced Material Research (CAMR) to Kurukshetra University. Authors are indebted to Dr. Dinakar Kanjilal, Raja Ramana Fellow, and Dr. Sundeep Chopra, Inter University Accelerator Centre (IUAC), New Delhi for fruitful discussions. We are thankful to Dr. P. C. Kalsi (Retd.), Scientist-H, Radiochemistry Division, BARC, Mumbai for useful comments and constructive insights.

Author details

Divya Gupta[1*], Rimpi Kumari[1], Amena Salim[2], Rahul Singhal[2] and Sanjeev Aggarwal[1]

1 Ion Beam Centre, Department of Physics, Kurukshetra University, Kurukshetra, India

2 Department of Physics, Malviya National Institute of Technology (MNIT), Jaipur, India

*Address all correspondence to: divyagupta2017@kuk.ac.in

References

[1] Zhao YP, Wang GC, Lu TM. Characterization of Amorphous and Crystalline Rough Surfaces: Principles and Applications. San Diego: Academic Press; 2001

[2] Bennett JM, Mattsson L. Introduction to Surface Roughness and Scattering. Washington, DC: Optical Society of America; 1989

[3] Eklund EA, Snyder EJ, Williams RS. Correlation from randomness: Quantitative analysis of ion-etched graphite surfaces using the scanning tunnelling microscope. Surface Science. 1993;**285**:157. DOI: 10.1016/0039-6028(93)90427-L

[4] Eklund EA, Snyder EJ, Williams RS. Submicron-scale surface roughening induced by ion bombardment. Physical Review Letters. 1991;**67**:1759

[5] Jacobs TDB, Junge T, Pastewka L. Quantitative characterization of surface topography using spectral analysis Surface Topography: Metrology and Properties. 2017;**5**:13001

[6] Gupta D, Chawla M, Singhal R, Aggarwal S. Nanoscale structural defects in oblique Ar^+ sputtered Si (111) surfaces. Scientific Reports. 2019; **9**:15531

[7] Wei Q, Lian J, Zhu S, Li W, Sun K, Wang L. Ordered nanocrystals on argon ion sputtered polymer films. Chemical Physics Letters. 2008;**452**:124-128. DOI: 10.1016/j.cplett.2007.12.053

[8] Slepicka P, Nedela O, Sajdl P, Kolska Z, Svorcík V. Polyethylene naphthalate as an excellent candidate for ripple nanopatterning. Applied Surface Science. 2013;**285**:885-892. DOI: 10.1016/j.apsusc.2013.09.007

[9] Facsko S, Dekorsy T, Koerdt C, Trappe C, Kurz H, Vogt A, et al. Formation of ordered nanoscale semiconductor dots by ion sputtering. Science. 1999;**285**:1551-1553

[10] Bradley RM, Harper JME. Theory of ripple topography induced by ion bombardment. Journal of Vacuum Science and Technology A. 1988;**6**: 2390-2395

[11] Moon MW, Han JH, Vaziri A, Her EK, Oh KH, Lee KR, et al. Nanoscale ripples on polymers created by focussed ion beam. Nanotechnology. 2009;**20**: 115301-115308. DOI: 10.1088/0957-4484/20/11/115301

[12] Vázquez L, Cubero AR, Lorenz K, Palomares FJ, Cuerno R. Journal of Physics: Condensed Matter. 2022;**34**: 333002. DOI: 10.1088/1361-648X/ac75a1

[13] Cuerno R, Kim JS. A perspective on nanoscale pattern formation at surfaces by ion beam irradiation. Journal of Applied Physics. 2020;**128**: 180902

[14] Valbusa U, Boragno C, Mongeot FB. Nanostructuring surfaces by ion sputtering. Journal of Physics: Condensed Matter. 2002;**14**:8153-8175

[15] Li WQ, Zhan XY, Song XY, Si SY, Chen R, Liu J, et al. A review of recent applications of ion beam techniques on nanomaterial surface modification: Design of nanostructures and energy harvesting. Small. 2019;**31**:901820. DOI: 10.1002/smll.201901820

[16] Gupta D, Kumari R, Umapathy GR, Singhal R, Ojha S, Sahoo PK, et al. Self-assembled nano-dots structures on Si (111) surfaces by oblique angle sputter-deposition. Nanotechnology. 2019;**30**:

385301-385312. DOI: 10.1088/1361-6528/ab273a

[17] Gupta D, Umapathy GR, Singhal R, Ojha S, Aggarwal S. Ripple patterns over oblique Ar+ sputtered SiC/Si(111) surfaces: Role of preferential sputtering. Materials Letters. 2019;**307**(4):131011. DOI: 10.1016/j.matlet.2021.131011

[18] Buatier de Mongeot F, Valbusa U. Applications of metal surfaces nanostructured by ion beam sputtering. Journal of Physics: Condensed Matter. 2009;**21**:224022. DOI: 10.1088/0953-8984/21/22/224022

[19] Goyal M, Aggarwal S, Sharma A, Ojha S. Surface structuring in polypropylene using Ar^+ beam sputtering: Pattern transition from ripples to dot nanostructures. Applied Surface Science. 2018;**439**:380-385. DOI: 10.1016/j.apsusc.2018.01.002

[20] Goyal M, Gupta D, Aggarwal S, Sharma A. Self-organized nanopatterning of polycarbonate surfaces by argon ion sputtering. Journal of Physics: Condensed Matter. 2018;**30**: 284002-284009. DOI: 10.1088/1361-648X/aac966

[21] Kumari R, Gupta D, Singhal R, Sharma A, Aggarwal S. Surface patterning of high density polyethylene by oblique argon ion irradiation. Journal of Applied Physics. 2019;**126**: 155303-155310. DOI: 10.1063/1.5116889

[22] Bradley RM. Theory of nanoscale ripple topographies produced by ion bombardment near the threshold for pattern formation. Physical Review E. 2020;**102**:012807. DOI: 10.1103/PhysRevE.102.01280

[23] Sigmund P. Theory of sputtering. I. Sputtering yield of amorphous and polycrystalline targets. Physics Review. 1969;**184**:383. DOI: 10.1103/PhysRev.184.383

[24] Gupta D, Aggarwal S, Sharma A, Kumar S, Chopra S. 200 kV ion accelerator Facility at Kurukshetra University, India. Materials Letters. 2022;**308**:131294

[25] Lee JH, Won J, Jeong H-C, Kim DH, Lee DW, Han J-M, et al. Physicochemical analysis of ion beam-induced surface modifications on polyethylene glycol films for liquid crystal alignment. Liquid Crystals. 2019;**46**(12):1799-1807. DOI: 10.1080/02678292.2019.1606351

[26] Gnaser H, Reuscher B, Zeuner A. Propagation of nanoscale ripples on ion-irradiated surfaces. Nuclear Instrument Methods B. 2012;**285**:142. DOI: 10.1016/j.nimb.2012.05.028

[27] Zeigler JF, Ziegler MD, Biersack JP. SRIM 2008.04 software package. Available online: http://www.srim.org

Section 3

Assembly Technology

Chapter 6

Mechanical Self-Assembly Technology for 2D Materials

Kai-Ming Hu and Wen-Ming Zhang

Abstract

Self-assembled mechanical instabilities can offer a new technology roadmap for micro/nanopatterns of two-dimensional (2D) materials, which depends on the deterministic regulation of mechanical instability-induced self-assemblies. However, due to atomic thinness and ultra-low bending stiffness, different types of non-designable and non-deterministic multimode coupling mechanical instabilities, such as multimode-coupled crumpling, chaotic thermal-fluctuation-induced rippling, and unpredictable wrinkling, are extremely easy to be triggered in 2D materials. The above mode-coupled instabilities make it exceedingly difficult to controllably self-assemble 2D nanocrystals into designed morphologies. In this chapters, we will introduce a novel micro/nanopatterning technology of 2D materials based on mechanical self-assemblies. Firstly, a post-curing transfer strategy is proposed to fabricate multiscale conformal wrinkle micro/nanostructures of 2D materials. Secondly, we report a deterministic self-assembly for programmable micro/nanopatterning technology of atomically thin 2D materials *via* constructing novel 2D materials/IML/substrate trilayer systems. Finally, based on the micro/nanopatterning technology of 2D materials, we proposed a new fabrication method for the flexible micro/nano-electronics of deterministically self-assembled 2D materials including three-dimensional (3D) tactile and gesture sensors. We fundamentally overcome the key problem of self-assembly manipulation from randomness to determinism mode by decoupling mono-mode mechanical instability, providing new opportunities for programmable micro/nanopatterns of 2D materials. Moreover, mechanical instability-driven micro/nanopatterning technology enables simpler fabrication methods of self-assembled electronics based on 2D materials.

Keywords: 2D nanomaterials, self-assembly, mechanical instability, nanopatterning, self-assembled electronics

1. Introduction

Mechanical self-assemblies in two-dimensional (2D) materials have aroused great scientific interest due to their significant promotion of basic understanding of special physical and chemical phenomena, spanning selective chemical reactivity, abnormal thickness fluctuations, fusion-fission behaviors, and anisotropic friction [1–7], as well as exciting applications such as strain sensors, photodetectors, anticounterfeiting, and energy storage [8–13]. Compared to three-dimensional (3D) bulk materials, the atomically thin and inherently planar 2D materials are more susceptible to being self-assembled into complex 3D architectures *via* rolling [14, 15], folding [16, 17], kirigami/origami [18–20], and rippling [21–23].

As an interesting self-assembly behavior, mechanical instability is expected to develop a micro/nanopatterning technology driven by deterministic mechanical self-assembly. However, 2D materials are very sensitive to external stimuli external stimuli due to the ultrathin thickness and exceptionally low-bending rigidity nature, such as thermal vibration [24, 25], high-temperature growth-induced lattice mismatch [26, 27], built-in tension [28, 29], van der Waals interactions [30], and compressive stress [31–33]. Therefore, it is very challenging to controllably assemble 2D materials into designed 3D morphologies, which often results in unavoidable restacking and aggregation. 2D materials are randomly corrugated in 3D *via* intrinsic rippling induced by thermal vibrations [34]. The abnormal mechanical, electronic, and thermal performances of ideal 2D materials can be greatly weakened, which tremendously increase the difficulty of manufacturing and transferring 2D nanomaterials and reduces the quality of materials [26]. In order to overcome the above shortcomings, extensive research has been conducted on the formation mechanism and suppression methods of spontaneous mechanical instabilities and non-deterministic and uncontrolled self-assemblies, resulting in ultra-flat or even strictly 2D materials [24, 26, 27, 35–38]. However, it is still very challenging but interesting to implement deterministic mechanical self-assemblies and allow better control over the self-assembly-driven micro/nanopatterns of 2D materials.

Mechanical instabilities in 2D materials can be utilized to introduce non-invasively tunable out-of-plane 3D topological deformations [39–41], further adjust charge transfer and carrier scattering [42, 43], and even create flat bands in 2D materials [44]. The key point to realizing deterministic self-assembly technology is to make mechanical instabilities in 2D materials controllable. Nevertheless, complex coupled multimodal mechanical instabilities, such as crumpling, rippling, wrinkling, folding, and buckling-induced delamination, can easily be triggered at the same time [27]. They result in the corresponding self-assembled morphologies suffer poor designability. The ripples caused by thermal fluctuations [24, 25] and the inherent random wrinkles generated during the growth and transfer process [26, 27] are random behaviors, making it difficult to predict and control the instability modes, scales, and directions of self-assembly morphologies [26]. Uniaxial strain can trigger quasi one-dimensional (1D) ordered self-assembled structures, but they are still uncontrollable multidirectional structures [31–33]. In addition, by coupling folding mode instabilities with wrinkle mode instabilities at an uncertain rate, crumpling is easily formed in 2D materials, which can lead to multidirectional and multiscale 3D self-assembly morphologies [31, 33]. The chaotic self-assembled structures induced by multimode coupling mechanical instabilities with uncontrollable scales and directions cannot precisely adjust the deformation of 3D structures in any way, which largely restrict the potential applications of self-assembled 2D materials. Consequently, decoupling multimode coupling mechanical instabilities and obtaining deterministic self-assembly with controlled structural configurations in 2D materials are crucial, but highly challenging.

2. Deterministic mode mechanical self-assembly

A complementary metal oxide semiconductor-compatible programmable tuning strategy has been developed to decouple uncontrollable multimode-coupled mechanical instabilities into controllable mode-decoupled mechanical instability, and to achieve deterministic mode self-assembly of 2D materials (**Figure 1A**). As shown in **Figure 1A, II**, the key point that realizes mode decoupling of mechanical instability in 2D nanosheets is to introduce designable intermediate multifunctional layers (IMLs) into bilayer systems. IMLs, including hydrogen silsesquioxane (HSQ) and poly (methyl methacrylate) (PMMA), play a variety of roles in generating and controlling deterministic self-assemblies driven by mono-mode mechanical instability, including deterministic mode, direction, and scale self-assemblies. First, IMLs work as carrier materials to avoid unwanted mode mechanical instability that triggers deterministic mode self-assembly in 2D materials during transfer. Second, IMLs work as a structural layer to enhance the effective bending stiffness of 2D materials/IML structures for self-assembly with deterministic dimensions. Third, IMLs also work as a functional layer and protective layer to manufacture programmable deterministic directional self-assembly patterns of 2D materials.

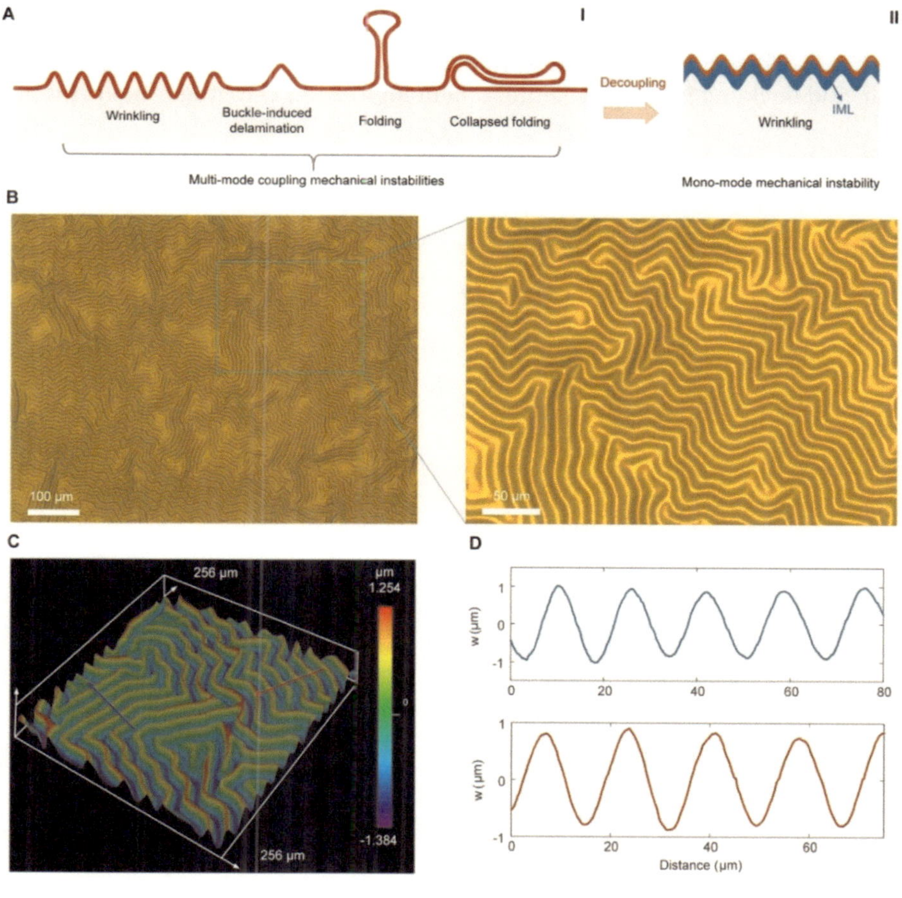

Figure 1.
Deterministic mode mechanical self-assembly in 2D materials driven by mode-decoupled mechanical instability. (A) Diagrammatic sketch of mechanical instabilities in tri-layered systems: (I) uncontrollable multimode-coupled mechanical instabilities in bilayer systems; (II) mode-decoupled mechanical instability triggered in 2D materials/IMLs/substrate three-layer systems with programmable IMLs. Microscopic morphologies of self-assembled MoS_2 induced by mono-mode instability: (B) microscopic image of mono-mode self-assembled morphologies with deterministic wavelength; (C, D) 3D and cross-sectional topology curves of laser scanning confocal microscopy (LSCM) images of self-assembled MoS_2.

Decoupled mono-mode instability and corresponding self-assembled patterns with deterministic wavelength of MoS_2 are observed in the designed three-layer systems in **Figure 1B** and **C**. Further, **Figure 1D** indicates that the deterministic self-assembly morphologies driven by mono-mode instability are mono-scale wrinkles with amplitude of 1.544 μm and wavelength of 16.392 μm. In stark contrast to the uncertain multiscale and multimode coupling instability, completely different decoupled single-mode instability and the resulting deterministic modes, namely single-scale self-assembly morphology, were observed in 2D materials. The decoupled mono-mode instabilities and the resulting deterministic mode, mono-scale self-assembly morphologies of 2D materials observed in **Figure 1B** and **C** are completely different from the non-deterministic multiscale and multimode coupling instabilities [21, 22, 39–41]. This is crucial for the success of self-assembly-based micro/nano-patterning technology in 2D nanocrystals. The more deterministic mode self-assembly morphologies of other 2D materials, including graphene, boron nitride (BN), tungsten disulfide (WS_2), and tungsten diselenide (WSe_2), are also uniform with highly deterministic mono-scale (**Figures 2–4**).

The results indicate that the mode-decoupled mechanical instability of 2D materials is achieved by constraining unwanted multimode instability (folding, collapsed folding, and buckling induced delamination in **Figure 1A, I**) to mono-mode instability

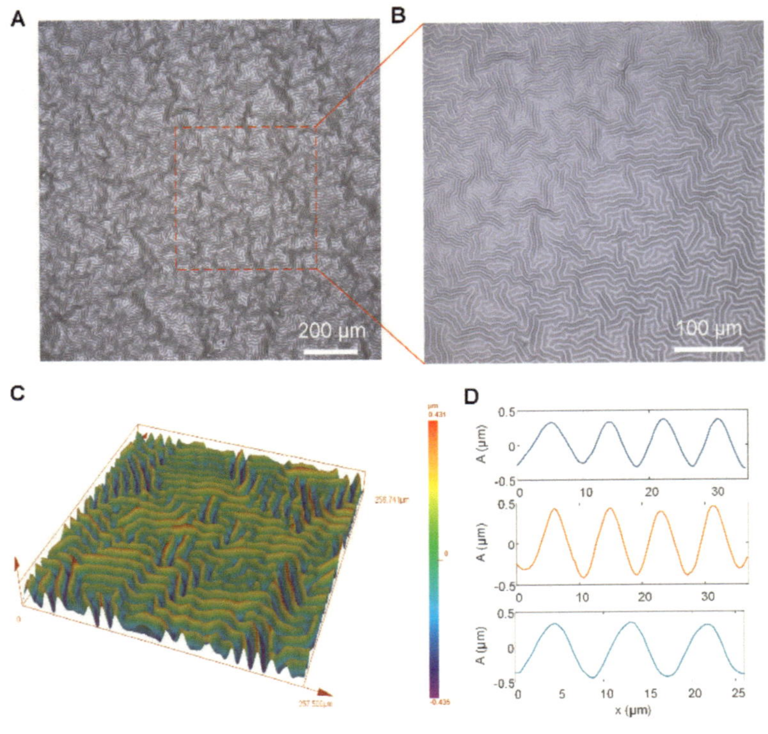

Figure 2.
Mono-mode instability-driven deterministic mode self-assembled MoS_2 in MoS_2/IML/beGr-PDMS trilayer systems: (A) microimages of large area, high-quality deterministic mode self-assembled structures of MoS_2 with 4% PMMA layer spin-coated at the speed of 6000 rpm; (B) the enlarged view of (A); (C) the 3D laser scanning confocal microscope (LSCM) images of the self-assembled structures of MoS_2; (D) 2D topological curves of the mono-mode instability-induced deterministic scale self-assembled structures at three different positions, which show the $\lambda = 8.470 \pm 0.130$ μm, $2A = 0.790 \pm 0.058$ μm for 4% PMMA under 6000 rpm.

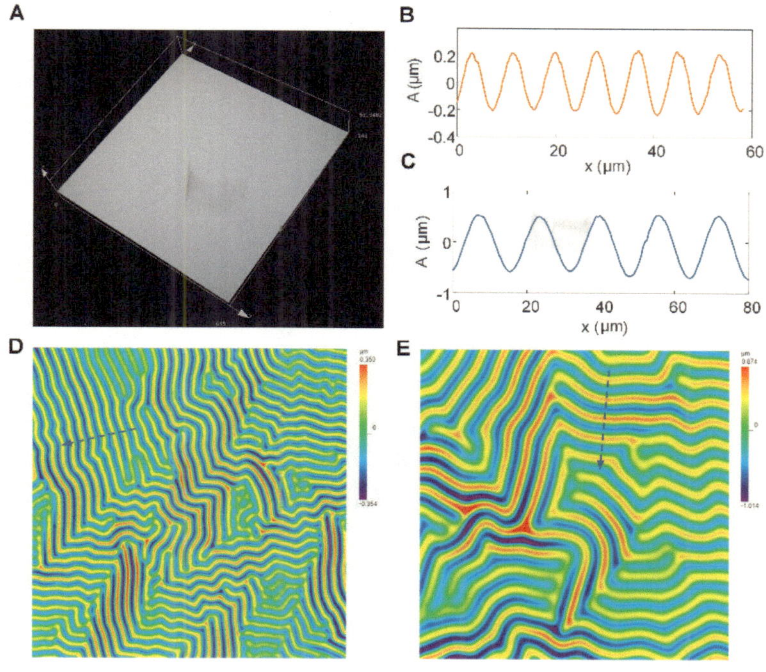

Figure 3.
Mono-mode instability-driven deterministic mode self-assembled WSe_2 in WSe_2/IML/beGr-PDMS trilayer systems: (A) microimages of transferred single-layer WSe_2 prepared by the low-temperature post-curing transfer method without the heat treatment triggering mono-mode instability; (B, C) the 2D topological curves of the deterministic mode self-assembled structures of WSe_2 under different thickness 4% PMMA layer controlled by different spin coating speeds 6000 rpm and 3000 rpm, respectively, where $\lambda = 7.770 \pm 0.400$ μm, $2A = 0.760 \pm 0.040$ μm for 4% PMMA at 6000 rpm, $\lambda = 15.500 \pm 0.550$ μm, $2A = 1.200 \pm 0.060$ μm for 4% PMMA at 3000 rpm; (D, E) 3D LCSM images of the self-assembled WSe_2 under different thickness 4% PMMA layer controlled by different spin coating speeds 6000 rpm and 3000 rpm, respectively.

(wrinkling in **Figure 1A, II**). As shown in **Figure 5**, the wavelength and amplitude of the self-assembly morphologies induced by mono-mode instability in three-layer systems can be given by

$$\lambda = 2\pi h_f \left(\overline{E}_f / 3\overline{E}_s\right)^{1/3} \tag{1}$$

$$A = h_f \sqrt{\varepsilon_0 / \varepsilon_c - 1} \tag{2}$$

where E_f and $h_f = h_{2D} + h_{IML}$ are the effective flexural modulus and thickness of 2D materials/IMLs surface films, respectively; ε_c is the critical strain that triggers the wrinkling mechanical instability. The feature parameters of orientation, amplitude, wavelength, and dynamic behaviors of the deterministic mechanical self-assembly patterns can be well adjusted, which is a key factor driving the progress of micro/nanopatterning technology based on mechanical self-assembly.

A new low-temperature post-curing transfer strategy is developed to achieve desirable and controllable mode decoupling mechanical instability in atomic thin 2D materials. The transfer strategy has three key points. It is extremely difficult to decouple and control the multimode-coupled mechanical instability of single-layer or even few layer 2D materials due to the extremely low bending stiffness [27, 39, 45]. Firstly, the well-designed IMLs we introduced to build 2D material/IML/substrate

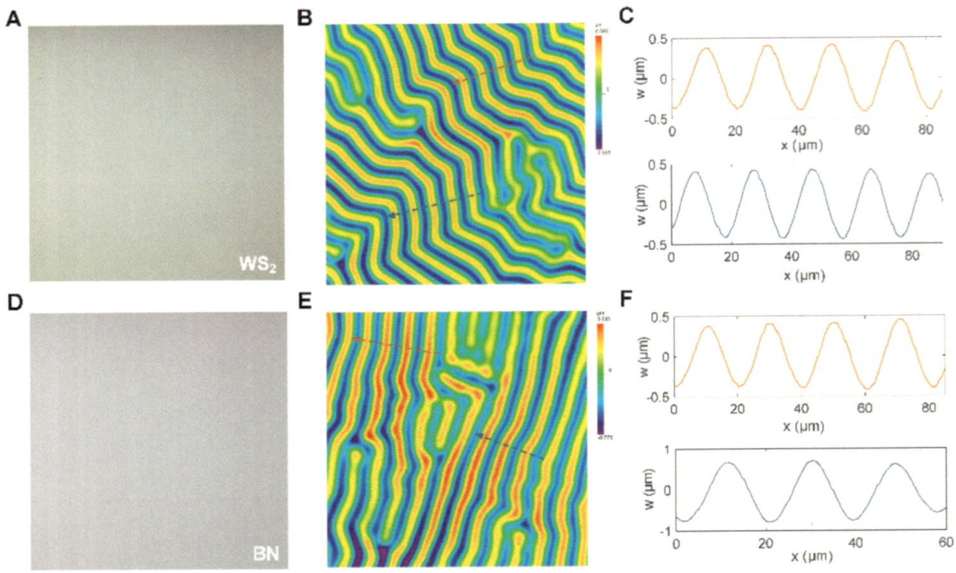

Figure 4.
Mono-mode instability-driven deterministic mode self-assembled WS_2 and BN in 2D materials/IML/beGr-PDMS trilayer systems: (A, D) microimages of transferred single-layer WS_2 and BN prepared by the low-temperature post-curing transfer method without the heat treatment triggering mono-mode instability; (B, E) 3D LCSM images of the deterministic mode self-assembled structures of WS_2 and BN under 4% PMMA layer controlled by spin-coating speed 3000 rpm; (C, F) 2D topological curves of the mono-mode instability-driven self-assembled structures of WS_2 and BN under 4% PMMA layer at spin-coating speeds 3000 rpm, respectively.

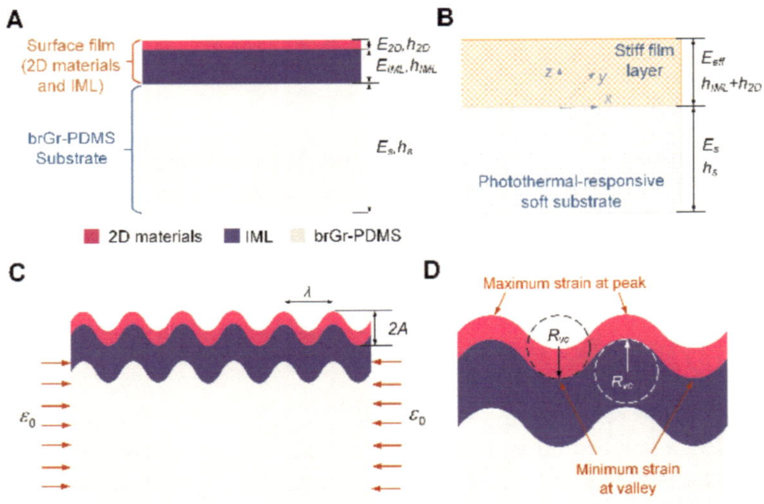

Figure 5.
Mechanical modeling for conformally self-assembled 2D materials in 2D materials/IML/beGr-PDMS trilayer system: (A) the trilayer systems with stiff surface film including 2D materials and IML, and photothermal-responsive soft brGr-PDMS substrate, where the thickness and Young's modulus of 2D nanosheets are h_{2D} and E_{2D}, the thickness and Young's modulus of IML are h_{IML} and E_{IML}, the thickness and Young's modulus of compliant substrates are h_s and E_s, respectively; (B) the modified core-shell model of surface elasticity of the effective surface film in the trilayer system, where the effective thickness $h_f = h_{2D} + h_{IML}$ and the effective Young's modulus of surface film E_{f}; (C) the self-assembled state of the trilayer system, where λ and A are the wavelength and amplitude of self-assembled structures, ε_0 is the initial strain; (D) schematic diagram of the maximum and minimum strains occurring at the valleys and crests of the structures.

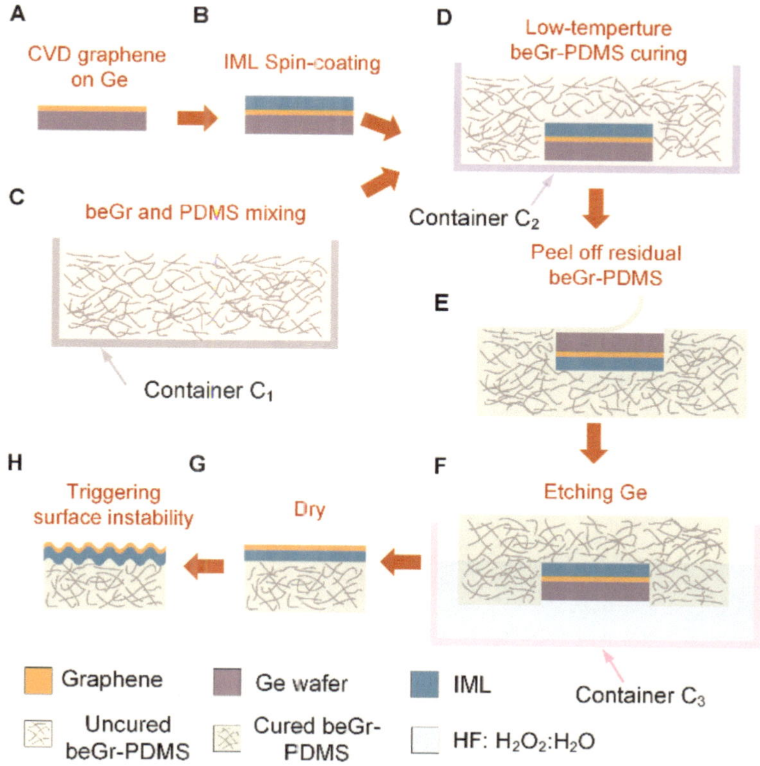

Figure 6.
Schematic diagrams of the low-temperature post-curing transfer process of graphene onto the PDMS: (A, B) a IML is spin-coated onto CVD-grown graphene/Ge surface, note that the IML layer works as a carrier material to prevent the unnecessary surface instabilities or fractures of graphene during transfer; (C) beGr-PDMS fabricated by mixing the uncured PDMS with bubble-exfoliated graphene (beGr) sheets; (D) the IML/graphene-Ge laminate is placed onto the bottom of container C_2 and the uncured beGr-PDMS is slowly poured onto the surface of the IML/graphene/substrate laminate, which is cured at low temperature (30°C) for 48 h to make beGr-PDMS fully cured; (E) the residual cured beGr-PDMS on the surface of Ge is peeled off; (F, G) the Ge layer is etched with HF: H_2O_2:H_2O solution for 4 h and the samples are rinsed with deionized water and dry on the heating plate to fully flatten graphene/IML film; (H) the mono-mode instability of graphene is triggered by the heating treatment (110° C) of the graphene/IML/beGr-PDMS system.

three-layer systems instead of two-layer 2D material/substrate systems to tackle the problem of extremely low flexural stiffness of 2D materials (**Figures 6**, **7**, and **8A–E**).

However, it also presents a new challenge that the interface strength among 2D materials, IMLs, and substrate is too low. The interfacial liquid introduced during the pre-curing transfer significantly lowers the interfacial adhesion energy between the pre-curing substrate and the 2D material, which cannot transfer stress (**Figure 9**). Note that the low-temperature post-curing transfer process of CVD-grown h-BN on Cu foil is similar to that of the CVD-grown graphene, the only difference is the grown substrate etching, in which the etchant is replaced by $FeCl_3$ solution. Unlike the conventional wet transfer methods of 2D materials [45], the carrier material in the graphene/IML laminate is not removed after they are transferred onto substrates, which is used as a structure layer to increase the bending stiffness of surface film.

Therefore, producing sufficient interfacial adhesion energy after transfer is the second key point, which can make the interface enable to subject to critical strain to generate the wanted mode mechanical instabilities. To improve interfacial strength between the 2D material and the substrate, we propose a post-curing transfer method

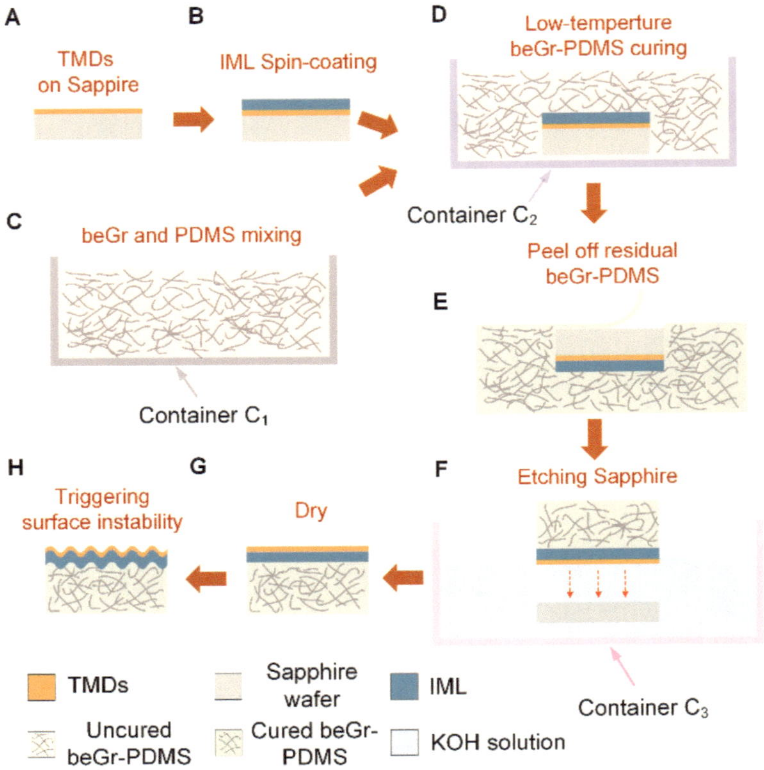

Figure 7.
Schematic diagrams of the low-temperature post-curing transfer process of 2D transition metal dichalcogenides (TMDs) materials, such as MoS_2, WS_2, and WSe_2, onto the PDMS: (A, B) the carrier material IML is spin-coated onto CVD-grown TMDS/sapphire surface; (C) beGr-PDMS fabricated by mixing the uncured PDMS with bubble-exfoliated graphene(beGr) sheets; (D) the IML-TMDs/sapphire composite layer is placed onto the bottom of container C_2 and the uncured beGr-PDMS is slowly poured onto the surface of IML-TMDs/sapphire laminate, which is cured at low temperature (30°C) for 48 h to make beGr-PDMS fully cured; (E) the residual cured beGr-PDMS on the surface of sapphire is peeled off; (F, G) the sapphire layer is slightly etched with KOH solution for 4 h and the samples are rinsed with deionized water and dry on the heating plate to fully flatten TMDs/IML film; (H) the mono-mode instability of 2D TMDs materials is triggered by the heating treatment (110°C) of the 2D TMDs/IML/beGr-PDMS trilayer system.

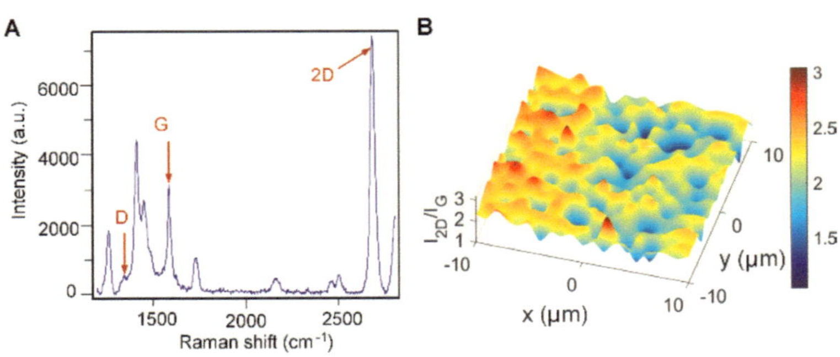

Figure 8.
Raman characterizations of graphene membranes transferred onto the PMMA-PDMS substrates: (A) D, G, 2D Raman, and other Raman spectra of graphene and PMMA in the samples; (B) 2D Raman results of I_{2D}/I_G for the samples after transfer.

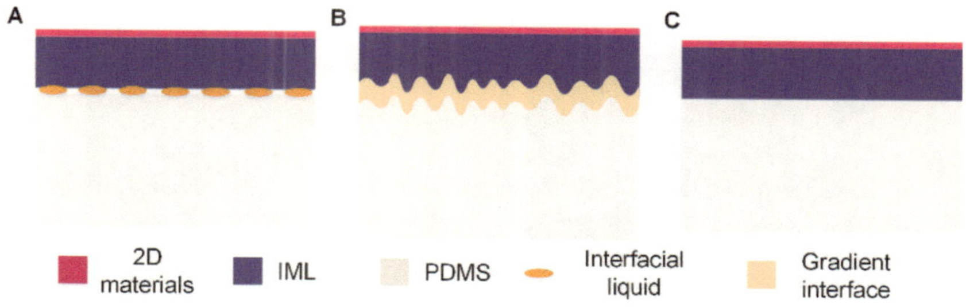

Figure 9.
Schematic diagrams of different interface conditions in film-substrate systems after 2D material transfer with different transfer methods: (A) the cross-sectional diagram of 2D materials/IML-substrate trilayer systems prepared by the pre-curing transfer method in [45], where interfacial liquid is trapped at the interface between IML and PDMS; (B) the cross-sectional diagram of 2D materials/IML-substrate trilayer systems prepared by the high temperature (70°C) post-curing transfer method, where no interfacial liquid is trapped at the interface between IML and PDMS, but high-temperature post-curing of PDMS can cause an gradient interface between IML and PDMS and multiscale self-assembled morphologies; (C) the cross-sectional diagram of 2D materials/IML-substrate trilayer systems prepared by the low-temperature (30°C) post-curing transfer method, where no interfacial liquid and gradient interface is existing at the interface between IML and PDMS, and mono-mode instability-induced deterministic scale self-assembled morphologies can be triggered.

for 2D materials, where the uncured beGr-PDMS is directly poured and then cured on the upper surface of IMLs (**Figure 10B** and **C**). However, the accelerated diffusion of uncured PDMS polymers by high temperatures can introduce an inevitable gradient interface between beGr-PDMS and IMLs during the post-curing process, which can lead to uncontrollable multiscale self-assemblies of 2D material in the three-layer system (**Figure 11**).

Therefore, the third point is avoiding the formation of diffusion-induced gradient interfaces between IMLs and beGr-PDMS substrate. To reduce polymer diffusion caused by Brownian motion, we cured the beGr-PDMS at low temperature (**Figure 10C**). To confirm this, we conducted cross-sectional characterization, which

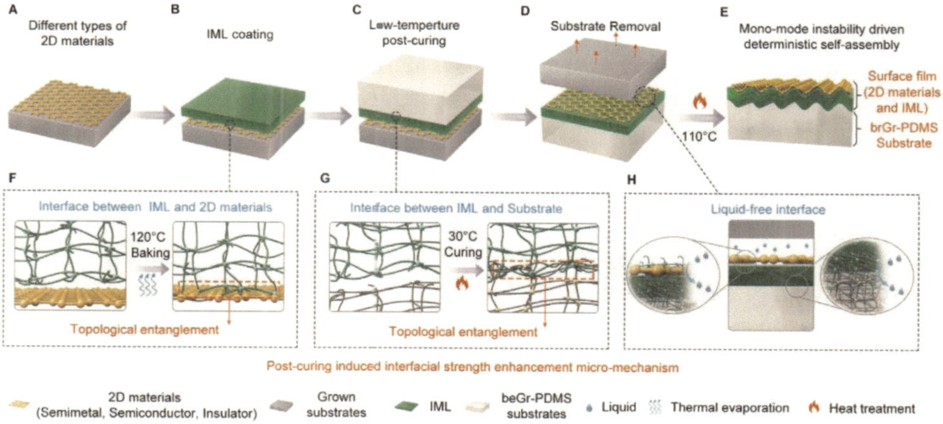

Figure 10.
Post-curing transfer strategy of 2D materials at low temperature: (A) different types of 2D materials grown by chemical vapor deposition (CVD); (B) IMLs spin-coating; (C) beGr-PDMS low-temperature curing; (D) substrate removal; (E) decoupled mono-mode mechanical instability and deterministic mode self-assembly triggered in trilayer systems with surface films (2D-materials and IMLs) and beGr-PDMS substrates. The mechanisms of interfacial strength enhancement caused by the post-curing transfer: (F, G) micro-mechanical interlocking in polymer networks; (H) liquid-free interfaces.

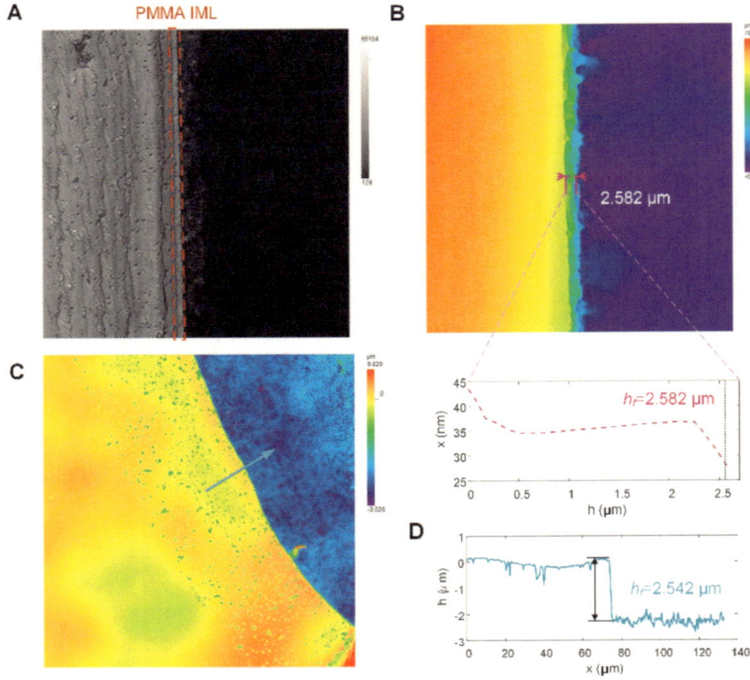

Figure 11.
(A) The optical micro-image of the cross sections of the samples prepared by the low-temperature posting-curing method, note that an obvious interface is observed in the sample and the red-dashed box is the PMMA IML; (B) 3D LSCM image of the cross sections of the 2D materials/PMMA-PDMS layered samples, where the thickness of 10% PMMA spin-coated by 1000 rmp is 2.582 μm. (C, D) 3D LSCM image of the top view and height profile in the 2D materials/PMMA-PDMS layered samples, where an obvious step with 2.542 μm is observed in the sample by peeling off 2D materials/PMMA surface films. Note that the height ($h_f = 2.582$ μm) of cross-sectional characterization is almost equal to the height ($h_f = 2.542$ μm) of top-view morphology characterization, which indicates that the obvious interface observed in cross-sectional characterization in (A, B) is the interface observed in (C, D).

indicates that a sharp interface between IMLs and beGr-PDMS substrate was observed instead of a gradient interface in the samples prepared by the low-temperature post-curing transfer method (**Figure 9**). In additions, Raman characterizations indicate that high-quality 2D materials with low defects are well transferred to target substrates (**Figures 8–14**). The results indicate that the low-temperature post-curing process can not only generate sufficient interface adhesion strength to withstand the critical interfacial stresses triggering mode-coupled mechanical instabilities, but also effectively avoid gradient interfaces induced by high temperatures-accelerated PDMS polymer diffusion.

We revealed the potential mechanism of the post-curing transfer-induced interfacial strength enhancement effect after solidification (**Figure 10F–H**). The enhanced interfacial strength between 2D materials and IMLs is caused by the thermal evaporation-driven topological entanglement of IMLs polymer chains during the high-temperature baking process (**Figure 10F**). Similarly, the interface strength enhancement between beGr-PDMS and IMLs is benefited by the topological entanglement between the IML and beGr-PDMS polymer network during the post-curing process (**Figure 10G**). The results indicate that the micro-mechanical interlocking induced by topological entanglement is the micro-mechanism of the interface strength

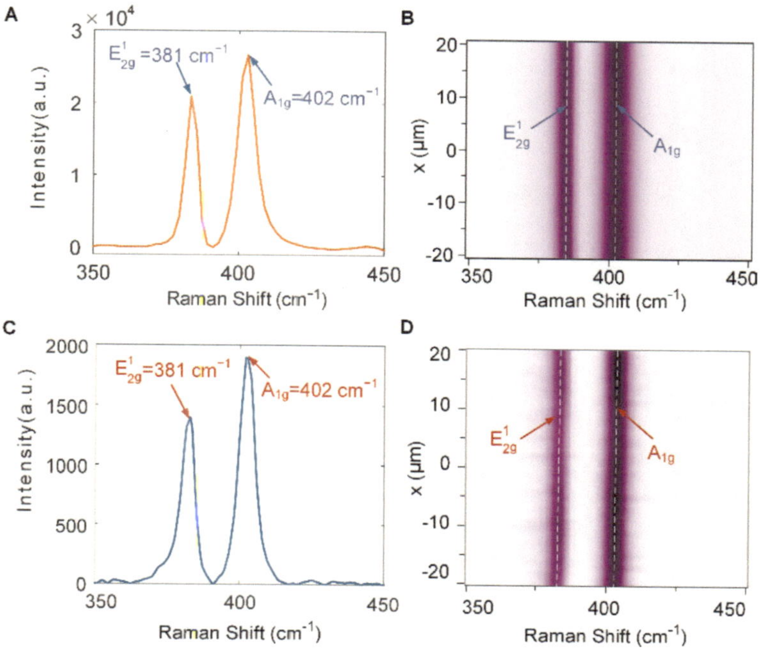

Figure 12.
Raman characterizations of single-layer MoS_2 (A, B) on SiO_2/Si substrates before transfer, and (C, D) on PMMA-PDMS substrates after transfer.

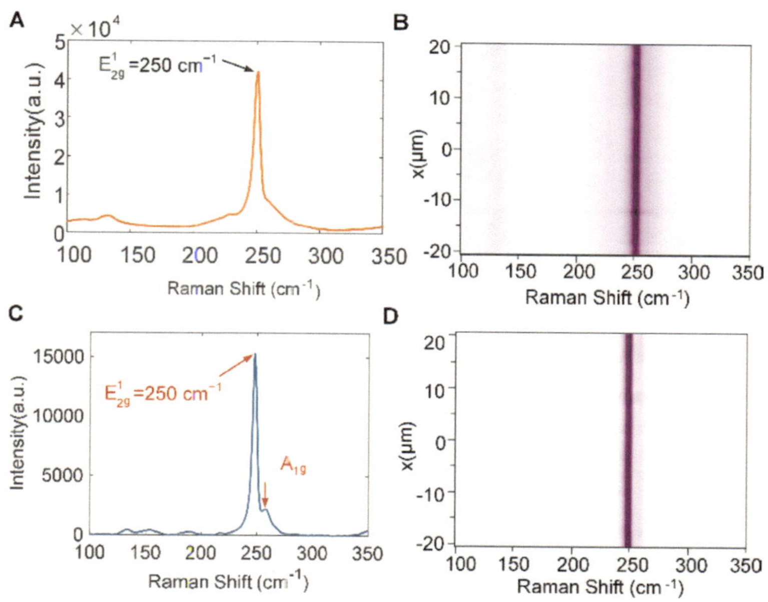

Figure 13.
Raman characterizations of single-layer WSe_2 (A, B) on sapphire substrates before transfer, and (C, D) on PMMA-PDMS substrates after transfer.

enhancement [46]. Furthermore, a sharp decrease in interface strength during substrate etching is effectively prevented due to the liquid-free interfaces between beGr-PDMS, IML, and 2D materials (**Figure 10D** and **H**).

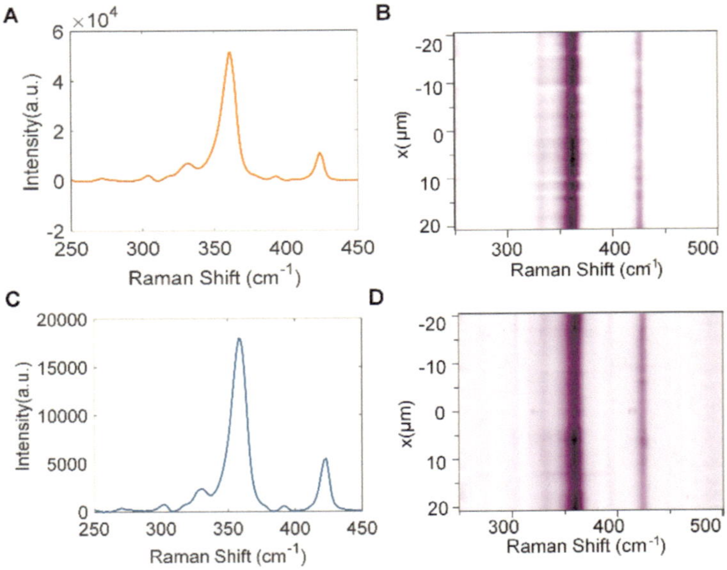

Figure 14.
Raman characterizations of single-layer WS_2 (A, B) on sapphire substrates before transfer, and (C, D) on PMMA-PDMS substrates after transfer. Note that Raman results in Figures S6–S8 indicate transferred single-layer 2D materials are highly uniform and of high quality.

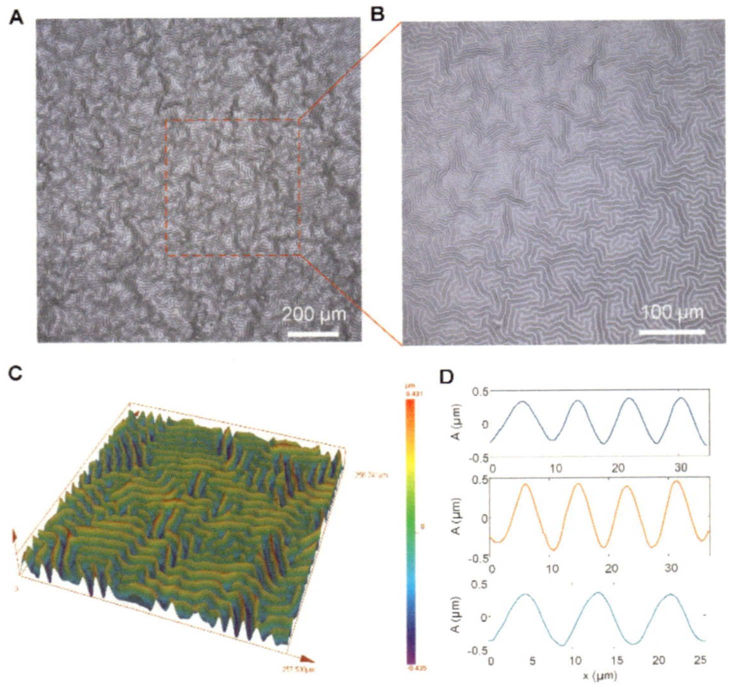

Figure 15.
Deterministic mode self-assembled MoS_2 in MoS_2/IML/beGr-PDMS trilayer systems: (A) microimages of large area, high-quality deterministic mode self-assembled structures of MoS_2 with 4% PMMA layer spin-coated at the speed of 6000 rpm; (B) the enlarged view of (A); (C) the 3D laser scanning confocal microscope (LSCM) images of the self-assembled structures of MoS_2; (D) 2D topological curves of the mono-mode instability-induced deterministic scale self-assembled structures at three different positions, which show the $\lambda = 8.470 \pm 0.130$ μm, $2A = 0.790 \pm 0.058$ μm for 4% PMMA under 6000 rpm.

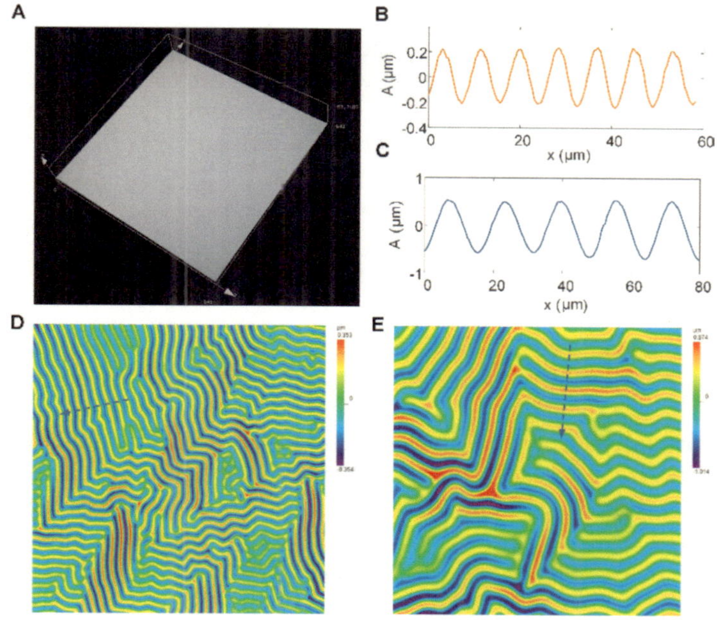

Figure 16.
Deterministic mode self-assembled WSe_2 in WSe_2/IML/beGr-PDMS trilayer systems: (A) microimages of transferred single-layer WSe_2 prepared by the low-temperature post-curing transfer method without the heat treatment triggering mono-mode instability; (B, C) the 2D topological curves of the deterministic mode self-assembled structures of WSe_2 under different thickness 4% PMMA layer controlled by different spin coating speeds 6000 rpm and 3000 rpm, respectively, where $\lambda = 7.770 \pm 0.400$ μm, $2A = 0.760 \pm 0.040$ μm for 4% PMMA at 6000 rpm, $\lambda = 15.500 \pm 0.550$ μm, $2A = 1.200 \pm 0.060$ μm for 4% PMMA at 3000 rpm; (D, E) 3D LCSM images of the self-assembled WSe_2 under different thickness 4% PMMA layer controlled by different spin coating speeds 6000 rpm and 3000 rpm, respectively.

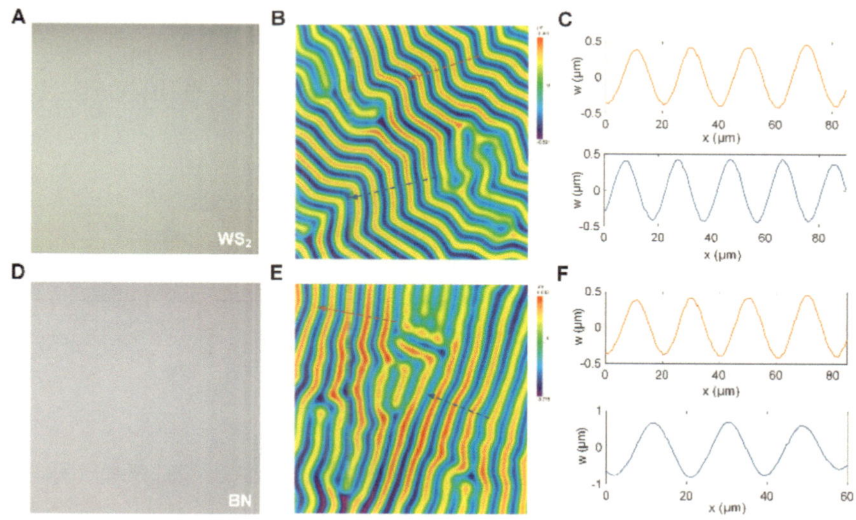

Figure 17.
Deterministic mode self-assembled WS_2 and BN in 2D materials/IML/beGr-PDMS trilayer systems: (A, D) microimages of transferred single-layer WS_2 and BN prepared by the low-temperature post-curing transfer method without the heat treatment triggering mono-mode instability; (B, E) 3D LCSM images of the deterministic mode self-assembled structures of WS_2 and BN under 4% PMMA layer controlled by spin-coating speed 3000 rpm; (C, F) 2D topological curves of the mono-mode instability-driven self-assembled structures of WS_2 and BN under 4% PMMA layer at spin-coating speeds 3000 rpm, respectively.

Moreover, more types of 2D materials grown by CVD, including insulator (h-BN), semi-conductors (2H-WS_2) and semi-metals (graphene, 1 T-WSe_2, 1 T-MoS_2), were transferred to prove the versatility of the transfer strategy (**Figure 10A**). As shown in **Figures 15–17**, the experimental data indicate that our transfer strategy has good universality and can be extended to other diverse 2D materials.

3. Cross-scale mechanical self-assembly

In order to achieve cross-scale self-assembly of 2D materials, IML was introduced as a structural layer to control the scale of deterministic mode self-assembly morphological features.

The IMLs are introduced to work as structural layers to control the characteristic scale of deterministic mode self-assembled patterns and achieve cross-scale self-assembly of 2D materials. The amplitude and wavelength of self-assembled micro/nanostructures in 2D materials can be controlled by the thickness of IMLs, judging from Eqs. (1)–(3). Furthermore, the thickness of IMLs can be programmatically regulated by spin-coating speeds$\omega_{spin}$and IMLs concentrations c_{IML}. As shown in **Figure 18A**, when the PMMA IMLs changes from 0.5% at 6000rmp to 10% at 1000rmp, the amplitude of self-assembled wrinkles increases from 8.5 ± 4.0nm to 5.203 ± 0.605μm. The wavelength has increased from 456 ± 46nm to 159.960 ± 8.500μm. The experimental results indicate that wavelength and amplitude can be effectively tuned by changing c_{IML} and ω_{spin}, and the tuning range is up to three orders of magnitude (**Tables 1** and 2; **Figures 19** and **20**). Judging from the mechanical optoelectronic coupling physical model of self-assembly in 2D materials/IMLs/substrate three-layer systems, cross-scale self-assembly is crucial for the micro/nano patterning technology based on self-assembly of 2D materials.

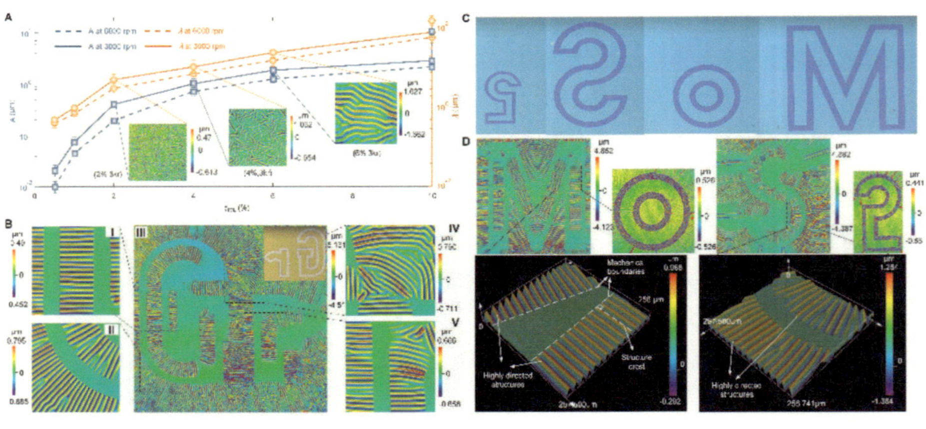

Figure 18.
Mode-coupled instability-induced cross-scale and directed self-assemblies of 2D materials in three-layer systems. (A) the amplitude and wavelength of micro/nanowrinkles in self-assembled 2D materials under different thickness IMLs, and the insets show the LSCM images of different scale self-assembled patterns in 2D materials. (B–D) 2D directed self-assemblies of 2D materials: (B) directed self-assembly of graphene with "gr"-shape; (C) EBL-defined mirrored "MoS_2" pattern before transfer; (D) 2D directed self-assembly of MoS_2 with "MoS_2" shape, in which the partial enlarged views are highly ordered MoS_2 with the structure crest orientations perpendicular to EBL-defined boundaries.

PMMA	0.5%, 6kr	1%, 6kr	2%, 6kr	4%, 6kr	6%, 6kr	10%, 6kr
λ (μm)	0.456 ± 0.046	0.760 ± 0.048	3.530 ± 0.380	8.125 ± 0.775	16.08 ± 0.775	59.97 ± 2.125
$2A$ (μm)	0.017 ± 0.008	0.051 ± 0.007	0.215 ± 0.021	0.765 ± 0.045	1.867 ± 0.085	2.163 ± 0.132

Table 1.
The tunable wavelength and amplitude of mono-mode instability-induced conformal self-assembled structures of 2D materials at spin-coating 6000 rpm with different IML concentrations.

PMMA	0.5%, 3kr	1%, 3kr	2%, 3kr	4%, 3kr	6%, 3kr	10%, 3kr	10%, 1kr
λ (μm)	0.566 ± 0.067	1.261 ± 0.266	5.700 ± 0.740	14.300 ± 1.850	25.220 ± 3.786	81.340 ± 5.560	159.960 ± 8.500
$2A$ (μm)	0.026 ± 0.005	0.085 ± 0.011	0.402 ± 0.042	1.020 ± 0.085	1.985 ± 0.136	2.866 ± 0.204	10.405 ± 1.210

Table 2.
The tunable wavelength and amplitude of mono-mode instability-induced conformal self-assembled structures of 2D materials at spin-coating 3000 rpm and 1000 rmp with different IML concentrations.

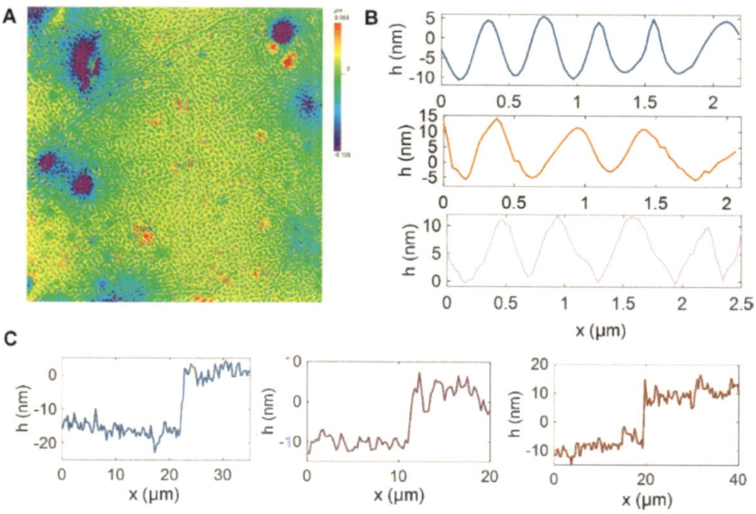

Figure 19.
Morphological characterizations of self-assembled nanostructures in the samples with IML thickness under 0.5% PMMA at spin-coating 3000 rpm: (A) LCSM images, (B) 2D topological curves of the mono-mode instability-induced self-assembled nano-structures. (C) Thickness characterization of IML under 0.5% PMMA at spin-coating 3000 rpm.

The wavelength and amplitude of mono-mode instability-induced conformal self-assembled 2D materials can be tuned by selecting different IML concentrations c_{IML} and spin-coating speeds ω_{spin}. The thickness of spin-coated IML layer can be given by

$$h_{IML} = k_{IML} c_{IML}^{n} \omega_{spin}^{-\gamma} \tag{3}$$

where k_{IML}, n, and γ are the coefficients depended on the IML type, viscosity value, and solvent vaporizability rate. As shown in Eq. (3), the thickness of IML h_{IML} can be tuned by the concentrations c_{IML} and spin-coating speeds ω_{spin} of IML.

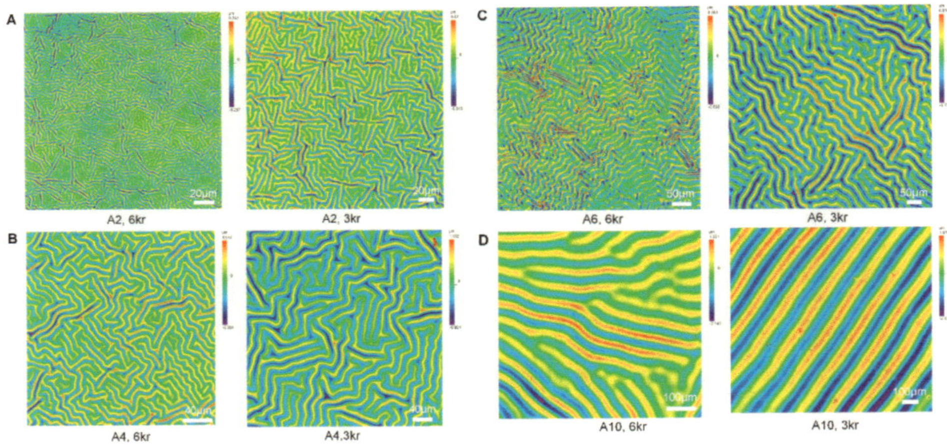

Figure 20.
LCSM morphological characterization of self-assembled structures in the samples with different thickness IMLs: (A) under 2% PMMA A2 at spin-coating 6000 rpm and 3000 rpm; (B) under 4% PMMA A4 at spin-coating 6000 rpm and 3000 rpm, (C) under 6% PMMA A6 at spin-coating 6000 and 3000 rpm; (D) under 10% PMMA A10 at spin-coating 6000 and 3000 rpm.

4. Directed mechanical self-assembly

As shown in Eqs. (1) and (2), the deformation orientation, anisotropic structural stiffness, and functionalization of 2D materials in three-layer systems of 2D materials/IMLs/substrate can be adjusted through directed self-assembly. Therefore, it is important for the applications of deterministic self-assembly to transform non-directed into directed self-assembly with designed directions, especially for self-assembly-driven directed micro/nanopatterning technology and 2D materials-based flexible devices. *Via* CMOS-compatible direct writing methods including electron beam lithography (EBL), the high-resolution designed mechanical boundaries in 2D material/IMLs/beGr-PDMS three-layer system are defined to realize directed self-assembly of 2D materials. It is worth noting that IMLs play a functional and protective role in directed self-assembly (**Figure 21**).

The above mechanical boundaries can be used to direct 2D materials self-assemble into ordered structures. **Figures 18B–D** and **22–24** show directed self-assembly morphologies of "MoS2" and "SJTU"-alphabetic-shaped MoS_2, and "Gr"-alphabetic-shaped graphene, in which IMLs, as electron beam resists, are used to programmatically fabricate ordered patterns. In the 2D oriented self-assembled graphene and MoS_2 nanosheets, the orientations of structural peaks are perpendicular to the EBL-defined boundaries (**Figure 18B** and **D**). Note that the EBL-defined "Gr" letter pattern was well transferred onto the target substrates by the low-temperature post-curing method without quality degradation. Moreover, the stress concentration-induced local self-assembly phenomenon can be harnessed to local strain engineering with large strains on demand by designing different curvature corners to control stress concentration and stress relaxation. We developed a revised shear-lag model reveals the mechanical mechanism of directed self-assembly structures induced by finite width soft constrained mechanical boundaries, indicating that mechanical boundaries can lead to asymmetric stress relaxation and asymmetric in-plane stress distribution.

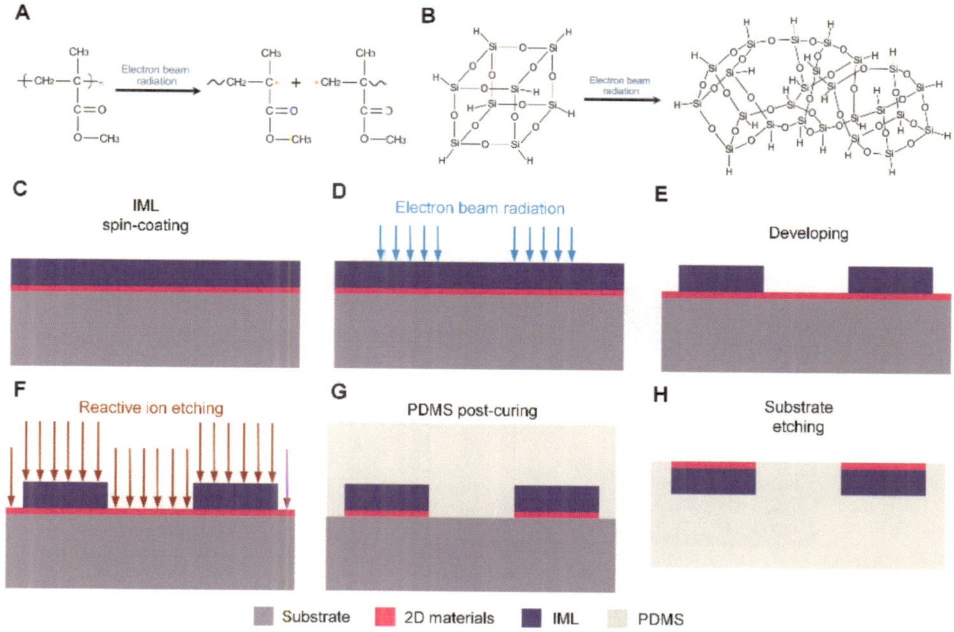

Figure 21.
IMLs-fueled programmable directed mechanical self-assembly of 2D materials: (A, B) photochemical reaction of IMLs (electron beam photoresist: PMMA and HSQ); (C–H) EBL-defined high-resolution mechanical boundaries, where (C, D) photochemical reaction of spin-coated IMLs is induced by the electron beam radiation, (E) the exposed IMLs are dissolved by the developer solution MIBK (Methyl Isobutyl Ketone): 2-proponol (IPA), (F) the 2D materials are removed via *reactive ion etching (RIE) and the IMLs work as the protective layer to protect the 2D materials covered by IMLs, (G, H) the post-curing of PDMS and the etching of the substrates.*

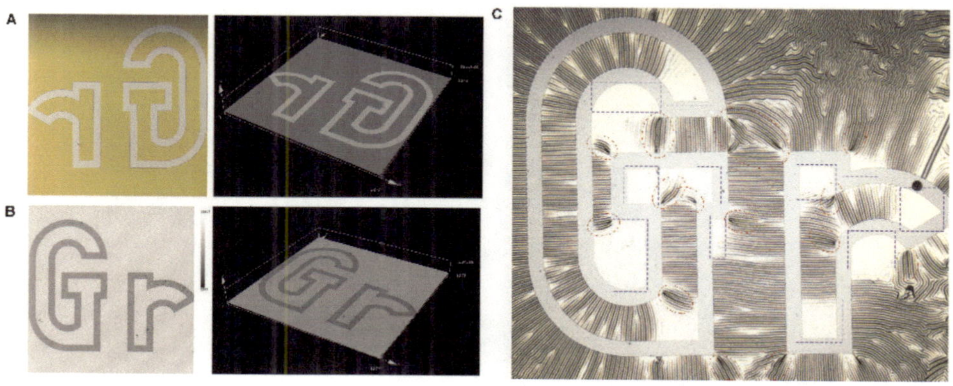

Figure 22.
"Gr" letter-shaped 2D directed self-assembled morphologies of graphene in layered systems with the IML under the condition of (4% PMMA, 3000 rmp): (A) mirrored "Gr" letters short for graphene before transfer, where the EBL-defined pattern is obtained by selectively dissolving the exposure area by the developer solution MIBK (Methyl Isobutyl Ketone): 2-proponol (IPA) for 45 s; (B) after transfer; (C) directed self-assembling structures of graphene with PMMA layer after heat treatment, where the red elliptic dashed box areas show the stress concentration-induced local mechanical instability with local increasing amplitudes around the inter corners (included angle < 180°), and the blue rectangle dashed box areas show the stress relaxation-induced local instability-free patterns around the outer corners (included angle > 180°).

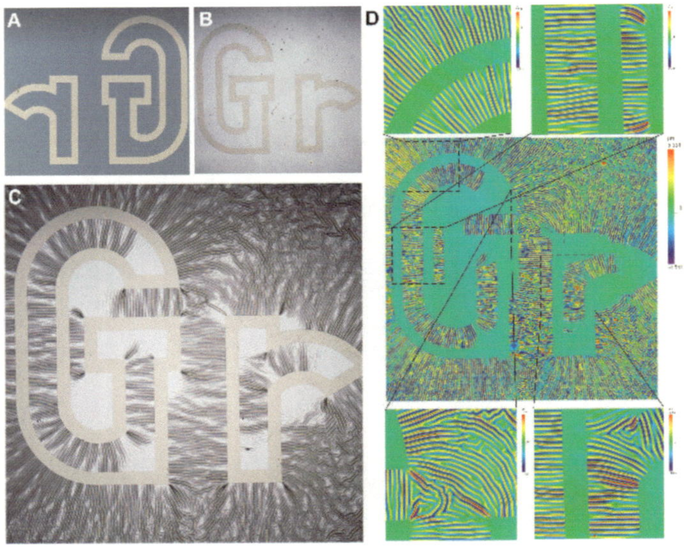

Figure 23.
"Gr" letter-shaped 2D directed self-assembled morphologies of graphene in layered systems with the IML under the condition of (4% PMMA, 6000 rmp): (A) mirrored "Gr" letters before transfer; (B) after transfer; (C) optical micro-images of directed self-assembling structures of graphene after heat treatment, similar with Figure S21, the stress concentration-induced local directed self-assembling structures around the inter corners and stress relaxation-induced local instability-free patterns around the outer corners are also observed.

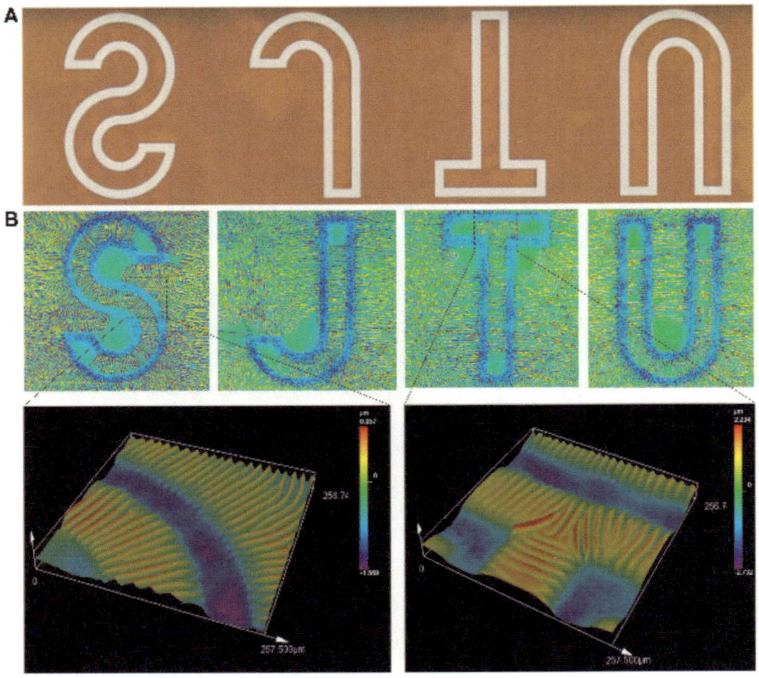

Figure 24.
"Gr" letter-shaped 2D directed self-assembled morphologies of MoS_2 in layered systems: (A) EBL-defined mirrored "SJTU" letters short for Shanghai Jiao Tong University before transfer and (B) LCSM images of 2D directed self-assembling structures after transfer and heat treatment, where the partial enlargement maps show the detailed directed self-assembling morphologies of "S" and "T" letters.

5. Dynamic mechanical self-assembly

In addition to the quasi-static above-mentioned cross-scale and directed self-assemblies, the dynamic self-assembly of 2D materials can be acquired through effective dynamic adjustment of mode-coupled instability. *Via* an interesting experimental design, a non-contact dynamic adjustment strategy is proposed to achieve *in situ*, real-time dynamic operation and synchronous self-assembly observation (**Figure 25A**). Firstly, graphene/conjugated polymer composite (beGr-PDMS) is constructed by mixing beGr [46] into PDMS. Rapid temperature changes and the resulting photothermal forces can be generated by near-infrared (NIR) irradiation of the beGr-PDMS substrate. Interestingly, we use multilayer 2D materials from the substrate to dynamically tune the self-assembly of single-layer 2D materials in the surface films. In addition, a reflector was designed to reflect and allow near-infrared light to enter the sample from the back of the substrate and to improve the incidence rate (**Figure 25A**).

During the dynamic operation of different NIR ON/OFF switches, the wavelength remains almost unchanged, indicating that the self-assembly morphology of 2D materials remains unchanged in the self-assembly mode and high mode-scale characteristics (**Figure 25B**). As indicated in **Figure 25B**, the amplitude of the self-assembled patterns can be dynamically adjusted by NIR light. When t = 0 s, $2A = 1.549 \pm 0.059$ μm; when

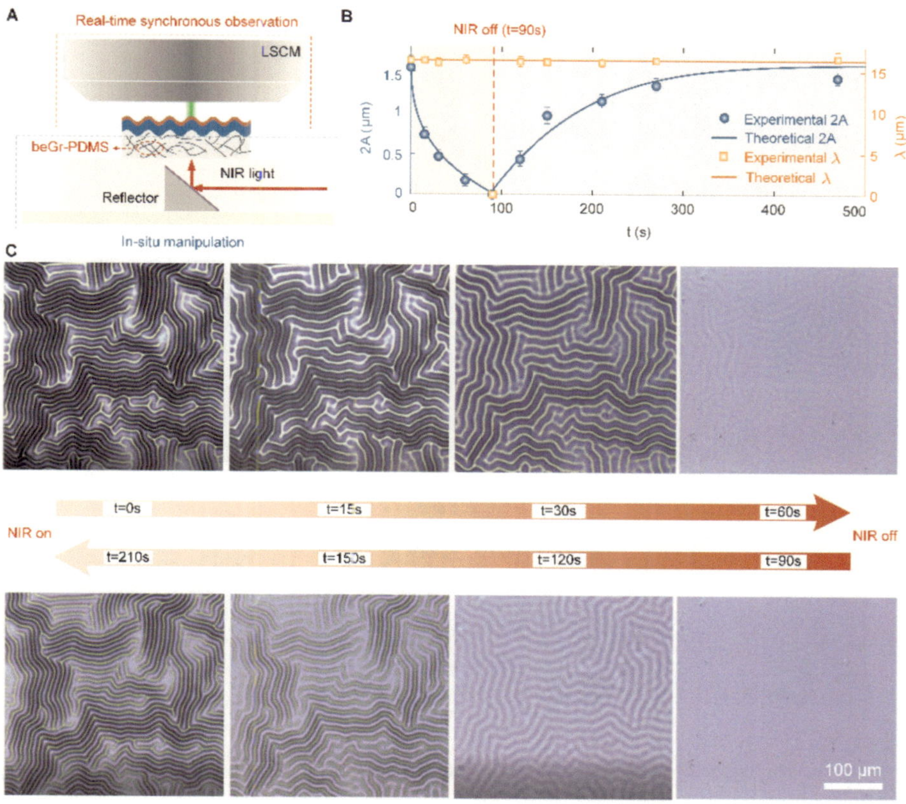

Figure 25.
Dynamic self-assembly of 2D materials in three-layer systems of 2D materials/IMLs/beGr-PDMS: (A) a platform for in situ *dynamic operation and real-time synchronous self-assembly morphology characterization; (B) when NIR is turned off at 90s, the time-varying wavelength and amplitude of the self-assembled structures; (C) NIR-controlled dynamic self-assembly.*

$t = 90$ s, $2A = 0$ μm. It is indicated that the self-assembled structures have been completely erased (**Figure 25C**). After NIR is turned off at $t = 470$s, wrinkles gradually recover and $2A = 1.417 \pm 0.070$ μm. We developed a thermodynamic model to study the dynamic manipulation mechanism and reveal the laws of dynamic self-assembly evolution of 2D materials in three-layered systems as follows

$$\Delta T(t) = \begin{cases} \kappa_1 t^{\alpha_1} + T_r & t \le t_{off} \\ T_r + (T_m - T_r)e^{-\alpha_2(t - t_{off})} & t > t_{off} \end{cases} \tag{4}$$

$$\sigma_0(t) = \frac{E_f(\alpha_s - \alpha_f)\Delta T(t)}{(1 - \nu_f)} \tag{5}$$

$$A(t) = h_f \sqrt{\sigma_0(t)/\sigma_c - 1} \tag{6}$$

where $\kappa_1, \alpha_1, \alpha_2$ are the constants related to photothermal energy conversion efficiency; $T_r, T_m = \kappa_1 t_{off}^{\alpha_1} + T_r$ are the room and highest temperatures; $A(t)$ is the time-varying amplitude of self-assembled structures; $\sigma_0(t)$ and $\Delta T(t)$ are the photothermal force and temperature difference controlling by NIR light; α_s, α_f are the thermal expansion coefficients of substrate and stiff thin film, respectively.

As shown in Eqs. (4)–(6) and **Figures 26** and **27**, the time-varying amplitude of self-assembled structures can be tuned by the time-varying photothermal force and

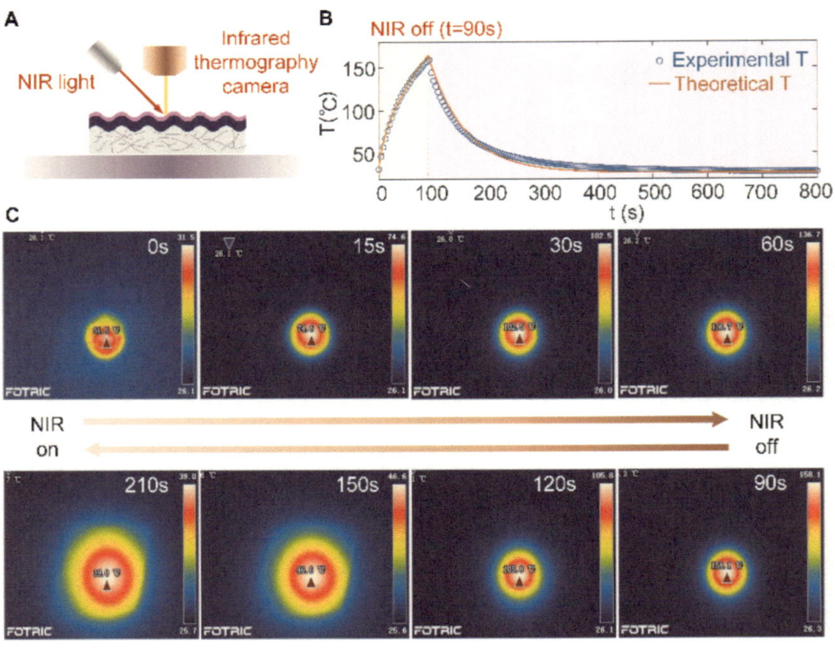

Figure 26.
Photothermal effect of multilayer graphene/conjugated polymer composites beGr-PDMS. (A) Schematic illustration of in situ *surface temperature measuring platform of samples by using an infrared thermography camera during NIR ON/OFF switch. (B) the variation curve of surface temperature with respect to irradiation time when NIR is off at 90 s, which can be depicted by piecewise function from Eq. (4) that well agrees with the experimental date. (C) the corresponding temperature distribution nephograms when the sample is illuminated by NIR light (808 nm, 2.0 W/cm^2) for 90 s. note that the beGr-PDMS heats up rapidly upon NIR irradiation when $t \le t_{off}$, which indicates that the high surface-area-to-mass ratio of beGr can result in high photothermal energy conversion of the beGr-PDMS.*

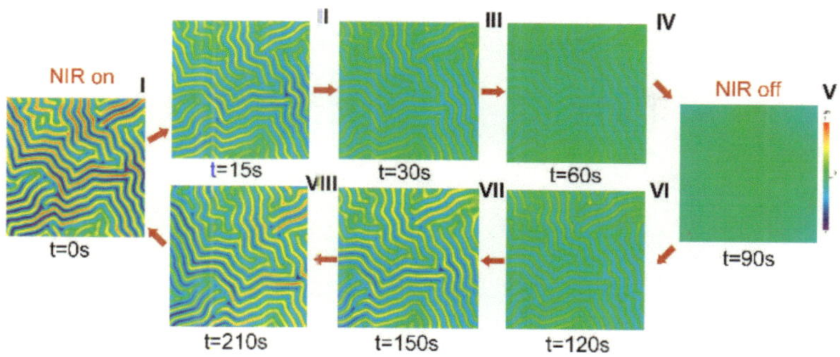

Figure 27.
Reversible dynamic evolution of self-assembled graphene when NIR is off at 90 s, where the IML (PMMA layer) is prepared by spin-coating 4% PMMA at 3000 rmp onto graphene surface and the power density of NIR is 2.0 W/cm². The amplitude of self-assembled structures decreases after the NIR ON, and the self-assembled structures are thoroughly erased (A = 0) when t = 90 s (from I to V). The amplitude of self-assembled structures increases after the NIR OFF and the self-assembled structures gradually reappear (from V to I).

temperature difference. The ratio of beGr, and power density and incidence rate of NIR can be used to modulated the photothermal response speed of beGr-PDMS.

The photothermal performance of NIR-responsive beGr-PDMS can be well described by Eq. (4) As shown in **Figure 25**, beGr-PDMS heats rapidly under near-infrared radiation, indicating a high conversion rate of photothermal energy. After the NIR is closed $(t > t_{off})$, the sample cools rapidly and tends to room temperature, which meets Newton's law of cooling (**Figure 25B** and **C**). The theoretical dynamic evolution of self-assembled wrinkles estimated by Eqs. (5) and (6) is very consistent with the experimental data (**Figure 25B**). As indicated in Eq. (6), the dynamic adjustment of strain and optoelectronic properties of 2D materials can be achieved by non-contact dynamic self-assembly, which opens up a way for the development of programmable micro/nano-patterning technology and dynamic tunable devices of 2D materials.

6. Self-assembled flexible devices

The deterministic self-assembled structures can bring some new unique properties into 2D materials, such as anisotropic piezoresistive, flexoelectric, piezoelectric, and mechanical optoelectronic coupling effects of 2D directed self-assembled 2D materials (**Figure 18B–D**). Moreover, the dynamic self-assembly can be used to realize the *in situ* dynamically tunable optoelectronic characteristics (**Figure 25**). The above new characteristics of controllable deterministic self-assembly can offer unprecedented feasibility opportunities for the self-assembled electronic devices of 2D materials.

In this work, as shown in **Figure 28**, a novel concept is proposed to manufacture deterministic self-assembled flexible electronic devices of 2D materials driven by deterministic mono-mode instability. A single self-assembled structure can be considered as an independently adjustable self-assembled electronic component, such as a transistor, a capacitor, a resistor, or a converter (**Figure 28A**). The independently adjustable electronic component can be assembled into complex circuits and interact with the external environment *via* self-assembled patterns of 2D materials. It is worth noting that self-assembled electronic components can be post-programmed through a determined mode circuit connection to form an adjustable self-assembled circuit

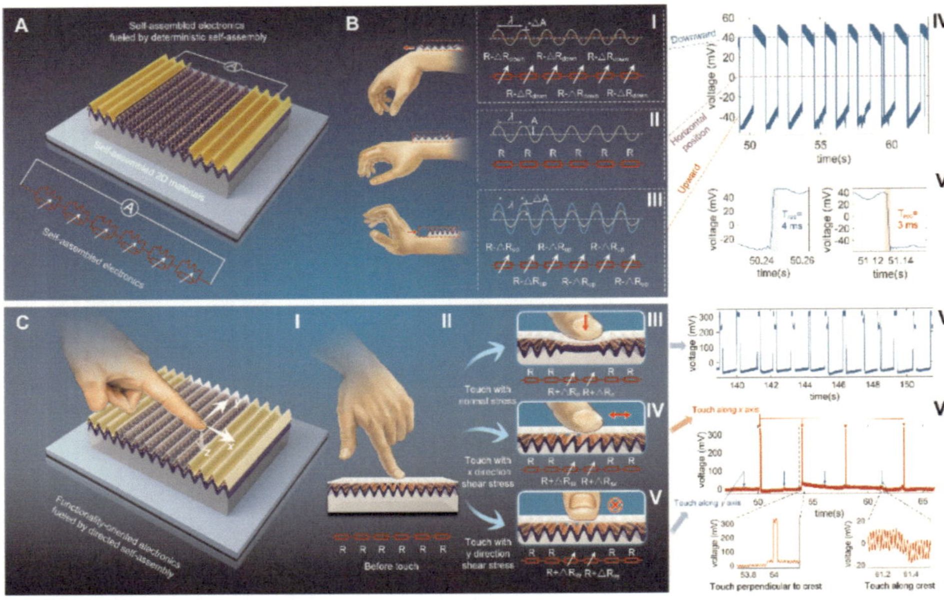

Figure 28.
Deterministically self-assembled flexible electronic devices: (A) schematics of self-assembled electronics; (B) flexible and wearable self-assembled electronics for gesture detection; (C) 3D anisotropic tactile sensors for normal and shear force discrimination.

network, such as resistors that can be linked in parallel, in series, or series parallel (**Figure 29**). In stark contrast, the 2D disordered multiscale self-assembly structures (**Figure 29A**) can lead to a non-deterministic connection network of random resistors (**Figure 29D**). The key to the self-assembled electronics of 2D materials driven by mono-mode instability is the deterministic self-assembly and circuit connections between electronic components.

As shown in **Figure 28B**, an ultra-fast response wearable flexible sensor for gesture detection was manufactured based on the deterministic self-assembly of 2D materials in 2D material/IML/substrate three-layer systems. When the hand is in the initial horizontal position, the sensor is connected to the arm near the wrist (**Figure 28B**). The amplitude and wavelength of initial self-assembled wrinkles are λ and A, and the initial resistance of the corresponding self-assembled resistor is R, respectively. The sensor is stretched when the wrist is rotated downward by 30 degrees, and the amplitude of self-assembled wrinkles decreases by $\triangle A$, the resistance of the self-assembled resistor also increases by $\triangle R_{down}$ and the voltage of the sensor increases by 40 mV (**Figure 28B, I–IV**). Furthermore, the sensor is compressed when the wrist rotates upward 30 degrees, resulting in a decrease in the amplitude of the self-assembled structures $\triangle A$, in the voltage and resistance of the sensor 40 mV and $N\triangle R_{up}$, where N is the number of the structures, respectively. The sensor has an extremely fast response speed and also is sensitive to wrist movements, where the response and recovery times $T_{rec} = 4\text{ms}$, $T_{rec} = 3\text{ms}$, respectively (**Figure 30**).

Moreover, by utilizing directed self-assembly of 2D materials in 2D material/IMLs/substrate three-layer systems, we fabricated a new functional oriented flexible electronic device (**Figure 28C**) [47]. A 3D anisotropic tactile sensor with ultra-fast response is presented by employing the 1D directed self-assembly of 2D materials to effectively identify normal and shear forces. The 1D ordered self-assembled structure

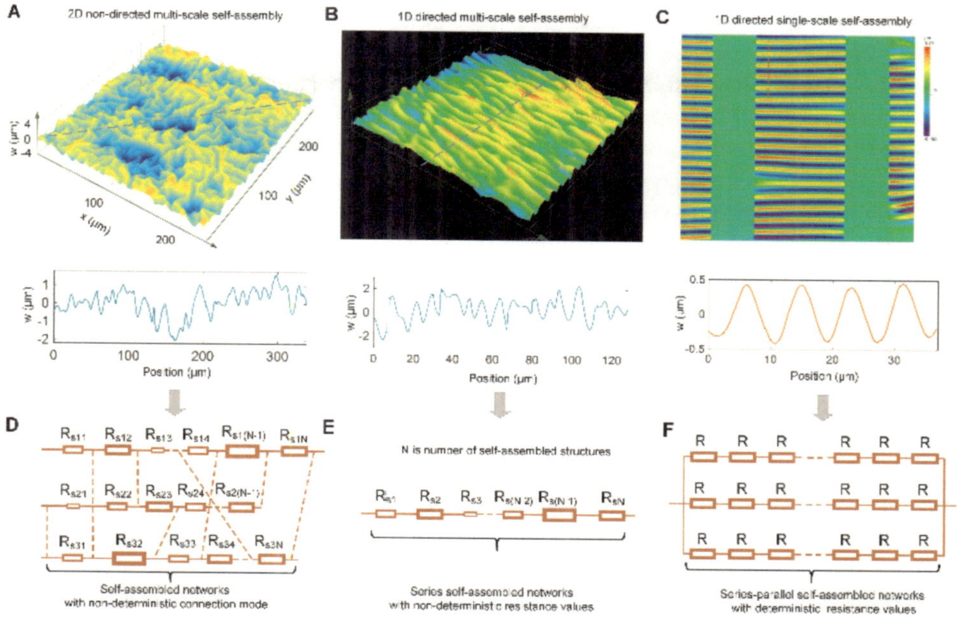

Figure 29.
The new concept of deterministically self-assembled micro/nano-electronics fueled by mode-decoupled mechanical instabilities in 2D materials. (A) Multimode coupling mechanical instabilities and 2D non-directed multiscale self-assembling structures; (B) multimode coupling mechanical instabilities and multiscale 1D directed self-assembling structures; (C) decoupled mono-mode mechanical instability and single-scale 1D directed self-assembling structures. The self-assembled micro/nano-electronics: (D) non-deterministic self-assembled networks of resistors, where resistors with random resistance values $R_{si}(i = 1, 2, 3, \ldots N)$ *are connected in a random manner, not deterministic series or parallel manner because 2D non-directed multiscale self-assembling structures are randomly aligned; (E) deterministic series self-assembled networks of resistors, where resistors with random resistance values are connected in deterministic series because 1D directed multiscale self-assembling structures are aligned in parallel; (F) deterministic series-parallel self-assembled networks of resistors, where resistors with deterministic resistance values* R *are deterministically connected in series and then in parallel because three stripes of 1D directed single-scale self-assembling structures are aligned in parallel.*

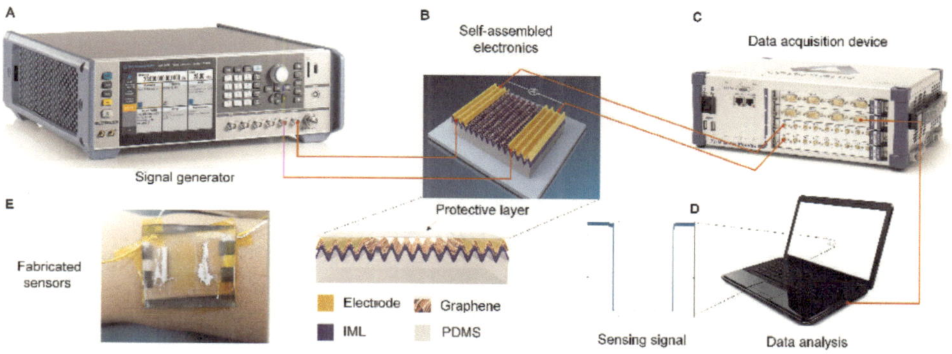

Figure 30.
The test platform of deterministically self-assembled electronics: (A) a signal generator is used to generate a constant current to drive electronic devices; (B) self-assembled electronics including a deterministically self-assembled 2D-materials/IML/substrate trilayer system, patterned Au electrodes, and a PDMS protective layer; (C) data acquisition device; (D) data analysis; and (E) fabricated sensors.

of 2D materials can work as sensing element, which is the key to the tactile sensing function of the anisotropic 3D tactile sensor. Each 1D ordered self-assembly structure can be equivalent to a resistor with the same resistance value, and the resistor is determined to be connected in parallel before being touched by fingers (**Figure 28C, II**). As shown in **Figure 28C, III–V**, the tactile sensor has three touch modes: normal force touch, x-direction shear force touch, and y-direction shear force contact. As indicated in **Figure 28C, VI**, the self-assembled structures are locally squeezed into a flat state, and the resistance ($\triangle R_n$) and voltage signal changes are observed due to the local structural deformation and piezoresistive effect of 2D materials. Due to the ultra-fast response characteristics of the sensor, it can accurately record the finger pressing time, where the response and recovery times T_{rec} = 4ms and T_{rec} = 3ms, respectively (**Figure 31**).

Furthermore, the sensor can also sense the shear force caused by finger sliding friction. Owing to 1D ordered self-assembly of 2D materials, the tactile sensors show anisotropic sensing characteristics. The self-assembled structure with a symmetrical sinusoidal shape undergoes significant deformation when the fingers slide along x direction, forming an asymmetric structure (**Figure 28C, IV**), with significant changes in resistance ($\triangle R_{sx}$) and sensing signal (**Figure 28C, VII**). However, the self-assembled structure undergoes very small deformation when the fingers slide along y direction (**Figure 28C, V**), without significant resistance changes ($\triangle R_{sy} \ll \triangle R_{sx}$) and sensing signals (**Figure 28C, VII**). The reason for that is the effective shear stiffness in x direction is much less than the stiffness in y direction of the 1D directed self-assembled structures. It is indicated that the sensor can effectively separate different direction shear forces. Moreover, the sensor can also distinguish between normal and shear forces. As shown in **Figures 31B** and **32B**, the *V-T* voltage signal curves of the sensors under one press-release and finger sliding cycle remain unchanged when the finger remains pressed. However, when the finger slides on the sensor, the voltage

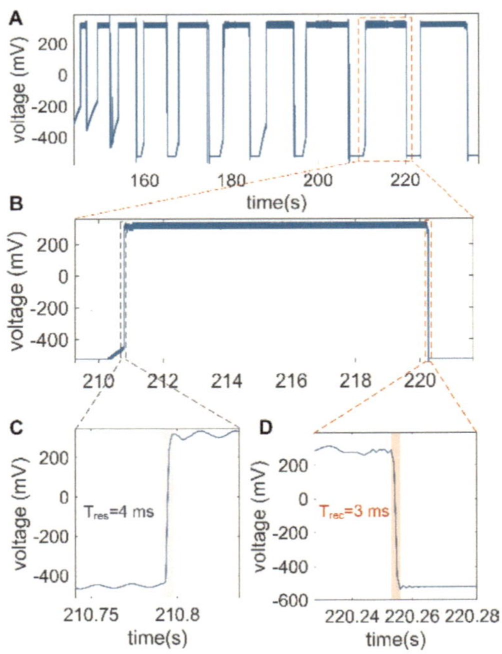

Figure 31.
Electric characterization of the ultra-fast responsive 3D tactile sensors in response to: (A) under finger press with increasing dwell times (from 1 s to 10s); (B) the enlarged view of V-T curves of the sensors under one press-release cycle; (C, D) V-T curves exhibiting ultra-fast responsive and recovery times, where the responsive time T_{res} *= 4ms and the recovery time* T_{rec} *= 3ms.*

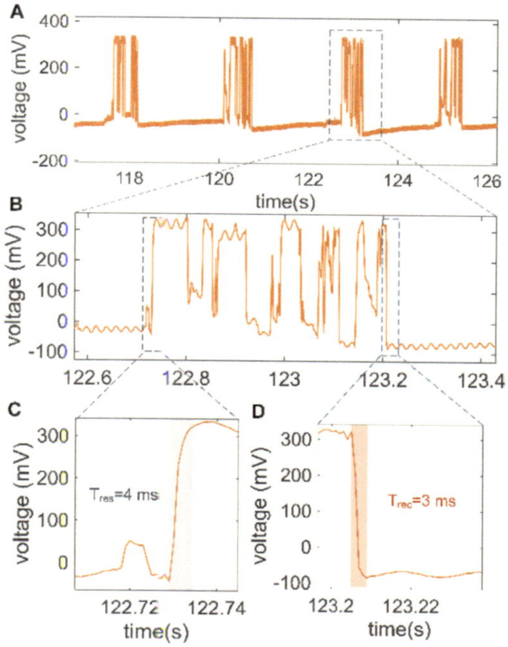

Figure 32.
Electric characterization of the ultra-fast responsive tactile sensors in response to: (A) under finger sliding motion with the shear force in x direction; (B) the enlarged view of V-T curves of the sensors under one finger sliding cycle; (C, D) V-T curves exhibiting ultra-fast responsive and recovery times, where $T_{res} = 4ms$ and $T_{rec} = 3ms$.

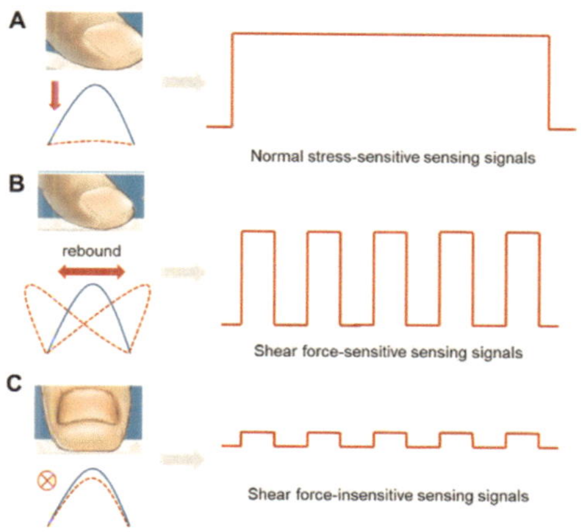

Figure 33.
The normal and shear force discrimination mechanical mechanism of anisotropic 3D tactile sensors: (A) the deformation mode and normal stress-sensitive sensing signals of the sensors under finger pressure; (B) the deformation mode and normal stress-sensitive sensing signals of the sensors when the fingers slide along x direction (the direction perpendicular to structural crest), where the sinusoidal self-assembled structure will rebound from left to right after the shear force is removed; (C) the deformation mode and normal stress-insensitive sensing signals of the sensors when the fingers slide along x direction (the direction parallel to structural crest).

signal repeatedly jumps up and down. The mechanical mechanism is that the self-assembled structures can repeatedly rebound after the shear force is eliminated (**Figure 33**). Therefore, the 3D anisotropic tactile sensors can effectively recognize the pressure and different direction shear forces from the fingers, owing to the directed self-assembled wrinkle-induced anisotropic piezoresistive effect of 2D materials. The deterministic self-assembly technology driven by mono-mode instability can provide a new and simpler manufacturing strategy for flexible electronic micro/nanodevices with inspiring functions.

7. Conclusions

We have fundamentally overcome the key problem of mechanical instability mode manipulation caused by the extremely low bending stiffness of atomically thin 2D crystals to realize the highly designable decoupled mono-mode mechanical instability and deterministic mode self-assembly of 2D materials. The key to decoupling multimode-coupled mechanical instabilities in ultra-thin 2D materials is to construct 2D material/IMLs-substrate three-layer systems by introducing a universal, programmable IMLs and micro-mechanical interlocking bonding.

The decoupled mono-mode instability of atomic thin 2D materials opens up a new opportunity for the determination of third-order cross-scale and directed self-assemblies in 2D nanocrystals. Interestingly, a non-contact *in situ* dynamic manipulation strategy was proposed for the post-tunable dynamic self-assembly of 2D materials. In addition, the above-mentioned programmable deterministic self-assembly endows 2D materials with new characteristics, such as anisotropic mechanical optoelectronic coupling and piezoresistive effects. A new concept is presented to manufacture deterministically self-assembled micro/nano-electronic devices based on 2D materials, such as ultra-fast response 3D anisotropic tactile sensors and wearable gesture sensors by using the above deterministic self-assembly inducing properties. The deterministic self-assembly driven by mono-mode instability can greatly promote CMOS-compatible micro/nanopatterning technology for 2D materials and deformable self-assembled electronic devices.

Acknowledgements

We strongly appreciate discussions with Yu-Long Chen, Shuai-Chen, Shu-Zheng Yan, Tian-Jiao Ma, Zhong-Ying Xue. We also thank Guang-Yu Zhang's group for providing 2D materials, Qi Sun for the EBL process, Gang Wang for providing beGr, and Tian-Tian Li for LCSM measurements. The authors would like to acknowledge the supports by the National Science Foundation (12121002, 12032015, 12172216), Science and Technology Innovation Action Plan of Shanghai (21190760100), the Program of Shanghai Academic/Technology Research Leader (19XD1421600), the State Key Laboratory of Mechanical System and Vibration (Grant No. MSVZD202105), and Double First-Class Construction project of Shanghai Jiaotong University (WH220402002).

Author details

Kai-Ming Hu* and Wen-Ming Zhang
State Key Laboratory of Mechanical System and Vibration, School of Mechanical Engineering, Shanghai Jiao Tong University, Shanghai, China

*Address all correspondence to: hukaiming@sjtu.edu.cn

References

[1] Meyer JC, Geim AK, Katsnelson MI, Novoselov KS, Booth TJ, Roth S. The structure of suspended graphene sheets. Nature. 2007;**446**(7131):60-63

[2] Xie S et al. Coherent, atomically thin transition-metal dichalcogenide superlattices with engineered strain. Science. 2018;**359**(6380):1131-1136

[3] Xue J et al. Scanning tunnelling microscopy and spectroscopy of ultra-flat graphene on hexagonal boron nitride. Nature Materials. 2011;**10**(4): 282-285

[4] Choi JS et al. Friction anisotropy–driven domain imaging on exfoliated monolayer graphene. Science. 2011; **333**(6042):607-610

[5] Chang D et al. Reversible fusion and fission of graphene oxide-based fibers. Science. 2021;**372**(6542):614-617. DOI: 10.1126/science.abb6640

[6] Deng S et al. Graphene wrinkles enable spatially defined chemistry. Nano Letters. 2019;**19**(8):5640-5646

[7] Sun P et al. Limits on gas impermeability of graphene. Nature. 2020;**579**(7798):229-232

[8] Kang P, Wang MC, Knapp PM, Nam S. Crumpled graphene photodetector with enhanced, strain-tunable, and wavelength-selective photoresponsivity. Advanced Materials. 2016;**28**(23):4639-4645

[9] Xiong P et al. Strain engineering of two-dimensional multilayered heterostructures for beyond-lithium-based rechargeable batteries. Nature Communications. 2020;**11**(1): 1-12. DOI: 10.1038/s41467-020-17014-w

[10] Wang Y et al. Super-elastic graphene ripples for flexible strain sensors. ACS Nano. 2011;**5**(5):3645-3650

[11] Thomas AV et al. Controlled crumpling of graphene oxide films for tunable optical transmittance. Advanced Materials. 2015;**27**(21): 3256-3265

[12] Jing L et al. Multigenerational crumpling of 2D materials for Anticounterfeiting patterns with deep learning authentication. Matter. 2020; **3**(6):2160-2180

[13] Dai C et al. Kirigami engineering of suspended graphene transducers. Nano Letters. 2022;**22**(13):5301-5306

[14] Zhao B et al. High-order superlattices by rolling up van der Waals heterostructures. Nature. 2021; **591**(7850):385-390

[15] Cui X et al. Rolling up transition metal dichalcogenide nanoscrolls via one drop of ethanol. Nature Communications. 2018;**9**(1):1-7

[16] Annett J, Cross GL. Self-assembly of graphene ribbons by spontaneous self-tearing and peeling from a substrate. Nature. 2016;**535**(7611):271-275

[17] Xu W et al. Ultrathin thermoresponsive self-folding 3D graphene. Science Advances. 2017;**3**(10): e1701084

[18] Blees MK et al. Graphene kirigami. Nature. 2015;**524**(7564):204-207

[19] Chen H et al. Atomically precise, custom-design origami graphene nanostructures. Science. 2019; **365**(6457):1036-1040

[20] Yang L-Z et al. Origami-controlled strain engineering of tunable flat bands and correlated states in folded graphene. Physical Review Materials. 2022;**6**(4): L041001

[21] Geringer V et al. Intrinsic and extrinsic corrugation of monolayer graphene deposited on SiO 2. Physical Review Letters. 2009;**102**(7):076102

[22] Hu J, Vanacore GM, Cepellotti A, Marzari N, Zewail AH. Rippling ultrafast dynamics of suspended 2D monolayers, graphene. Proceedings of the National Academy of Sciences of the USA. 2016; **113**(43):E6555-E6561

[23] Dobrik G et al. Large-area nanoengineering of graphene corrugations hosting visible-frequency graphene plasmons. Nature Nanotechnology. 2021;**17**(1):1-18

[24] Lui CH, Liu L, Mak KF, Flynn GW, Heinz TF. Ultraflat graphene. Nature. 2009;**462**(7271):339-341

[25] Fasolino A, Los JH, Katsnelson MI. Intrinsic ripples in graphene. Nature Materials. 2007;**6**(11):858-861. DOI: 10.1038/nmat2011

[26] Yuan G et al. Proton-assisted growth of ultra-flat graphene films. Nature. 2020;**577**(7789):204-208

[27] Wang M et al. Single-crystal, large-area, fold-free monolayer graphene. Nature. 2021;**596**(7873):519-524

[28] Calado V, Schneider G, Theulings A, Dekker C, Vandersypen L. Formation and control of wrinkles in graphene by the wedging transfer method. Applied Physics Letters. 2012;**101**(10):103116

[29] Zhao Y et al. Large-area transfer of two-dimensional materials free of cracks, contamination and wrinkles via controllable conformal contact. Nature Communications. 2022;**13**(1):1-10

[30] Ares P et al. Van der Waals interaction affects wrinkle formation in two-dimensional materials. Proceedings of the National Academy of Sciences of the USA. 2021;**118**(14):e2025870118

[31] Zang J et al. Multifunctionality and control of the crumpling and unfolding of large-area graphene. Nature Materials. 2013;**12**(4):321-325

[32] Lee W-K et al. Multiscale, hierarchical patterning of graphene by conformal wrinkling. Nano Letters. 2016;**16**(11):7121-7127

[33] Chen PY et al. Multiscale graphene topographies programmed by sequential mechanical deformation. Advanced Materials. 2016;**28**(18):3564-3571

[34] Novoselov KS, Geim A. The rise of graphene. Nature Materials. 2007;**6**(3): 183-191

[35] Xu P et al. Unusual ultra-low-frequency fluctuations in freestanding graphene. Nature Communications. 2014;**5**:1-7. DOI: 10.1038/ncomms4720

[36] Deng J et al. Epitaxial growth of ultraflat stanene with topological band inversion. Nature Materials. 2018;**17**(12): 1081-1086

[37] Deng B et al. Growth of Ultraflat graphene with greatly enhanced mechanical properties. Nano Letters. 2020;**20**(9):6798-6806

[38] Zheng F, Thi QH, Wong LW, Deng Q, Ly TH, Zhao J. Critical stable length in wrinkles of two-dimensional materials. ACS Nano. 2020;**14**(2): 2137-2144

[39] Castellanos-Gomez A et al. Local strain engineering in atomically thin MoS2. Nano Letters. 2013;**13**(11): 5361-5366

[40] Bao W et al. Controlled ripple texturing of suspended graphene and ultrathin graphite membranes. Nature Nanotechnology. 2009;**4**(9):562-566. DOI: 10.1038/nnano.2009.191

[41] Carbone MGP, Manikas AC, Souli I, Pavlou C, Galiotis C. Mosaic pattern formation in exfoliated graphene by mechanical deformation. Nature Communications. 2019;**10**(1):1-7

[42] Zhu W et al. Structure and electronic transport in graphene wrinkles. Nano Letters. 2012;**12**(7):3431-3436

[43] Pereira VM, Neto AC, Liang H, Mahadevan L. Geometry, mechanics, and electronics of singular structures and wrinkles in graphene. Physical Review Letters. 2010;**105**(15):156603

[44] Mao J et al. Evidence of flat bands and correlated states in buckled graphene superlattices. Nature. 2020; **584**(7820):215-220

[45] Hu KM et al. Tension-induced Raman enhancement of graphene membranes in the stretched state. Small. 2019;**15**(2):1804337

[46] He P et al. Kinetically enhanced bubble-exfoliation of graphite toward high-yield preparation of high-quality graphene. Chemistry of Materials. 2017; **29**(20):8578-8582

[47] Hu KM et al. Deterministically self-assembled 2D materials and electronics. Matter. 2023;**6**(5):1654-1668

Chapter 7

Stretching the Horizon of Nanosphere Lithography

Arnab Ganguly and Gobind Das

Abstract

Periodic nano-structured superlattice with symmetry breaking along the surface as well as thickness is of high importance in plasmonic research due to its immense potential in bio-medical and environmental sensor applications. The structures are rich with electromagnetic hotspots and anisotropic optical properties contributing to the sensitivity of the device. In nanotechnology, nanosphere lithography (NSL) is a promising lithography technique that is in the limelight of research for the last decade due to its simplicity, scalability, and cost-effectiveness. Despite many advantages, the technique is limited in its versatility when it comes to real application. Conventional NSL offers less scope for structural variation with the most common six-fold symmetric structure as the nanosphere mask self-assembles in a hexagonal close-pack fashion due to the very nature of this process. Research efforts have been made to overcome the bottleneck. The conventional NSL approach is modified in combination with other techniques. In this chapter, we discuss the evolution of the NSL technique to achieve maturity. The chapter emphasizes modern NSL formalism associated with techniques like multistep shadow mask deposition and anisotropic etching to achieve 2D and 3D nanostructures with controlled shape, size, inter-structural gap, aspect ratio, sharpness, and special and vertical symmetry. The chapter will serve as an effective knowledge base for further research development in plasmonics, magnonics, and spintronics.

Keywords: nanosphere lithography, shadow mask deposition, anisotropic etching, hotspot, sensors

1. Introduction

Nanosphere lithography (NSL) is a popular and cost-effective technique used to fabricate periodic patterns of nanoscale structures on a substrate using monodisperse polystyrene nanospheres as mask. In this technique, one or more layers of nanospheres are deposited onto a substrate followed by a thin monolayer or multilayer coating of target material such as metals and polymers. The material is deposited only in the spaces between the nanospheres. Later, the nanospheres are removed, leaving a periodic array of nanostructures on the substrate.

The NSL technique has been the research highlight for several decades due to its unbeatable advantages over other lithography techniques. In nanotechnology, conventional lithography (CL) is a popular tool used to fabricate various nanostructures [1–3]. However, in the case of photolithography, it is tricky to obtain nicely resolved structures in the sub-micron range. E-beam lithography has higher resolutions, but the technique fails when it comes to large-area patterning. Moreover, CL is not the right candidate for patterning 3D nanostructures (i.e., structures having asymmetry along thickness). Such a 3D structure can be patterned using a focused ion beam, but again it is limited to small-area patterning. In short, these techniques are expensive, and sensitive to cleanroom conditioning, hence, not suitable for large-scale industrial applications. On the contrary, NSL [4–8] is a package with all the qualities in it. It offers a simple alternative to realize large-area patterning of 2D and 3D nanostructures in a cost-effective manner and does not necessarily require special cleanroom conditioning. In fact, the entire process can be carried out using standard laboratory equipment and does not require much time to perform. NSL can produce high-resolution patterns with feature sizes down to a few nanometers, which is comparable to the electron beam lithography (EBL). It is a versatile technique and does not involve the use of harsh chemicals, making it a greener alternative compared to its peers.

Despite the advantages, the technique has some bottlenecks which are limiting the realization of this technique in modern technology to its full potential. The uniformity of nanosphere mask assembly is the key to successful patterning. However, the self-assembly process is stochastic and sensitive to various environmental parameters such as temperature, humidity, hydrophilicity, and pH, which can affect the quality and reproducibility of the patterns. Defects and dislocations present in the nanosphere mask are carried forward to the patterned structure. NSL is typically limited to flat substrates, such as glass or silicon wafers, which can be problematic for applications that require patterned features on non-planar or curved surfaces. Material compatibility of the substrate for NSL is an also issue. Another concern to the NSL is its limited control of structures. In other words, the technique cannot produce a wide range of nanostructures. Most conventionally, it is used to create triangular crystals with hexagonal symmetry as the technique is based on nanosphere masks which self assembles in hexagonal close-packed (HCP) order [8, 9].

Research efforts have been made to overcome these challenges. Recently, significant development has been made over the conventional NSL approach to achieve complex 2D/3D nanostructures over a large area. For instance, a symmetry breaking along the thickness through non-uniform etching has been observed by Darvill et al. [10]. Myint et al. [11] observed crystal structures from three-fold azimuthal symmetry deposition. The development enhances the appeal and relevance of the NSL technique in plasmonic sensors, magnonic memory, optoelectronics, and biomedical applications. This futuristic evolution of the NSL technique is called "stretching the horizon" in the title of this chapter. The chapter emphasizes modern NSL formalism associated with techniques like multistep shadow mask deposition and anisotropic etching in order to achieve 2D and 3D nanostructures with controlled shape, size, inter-structural gap, aspect ratio, sharpness, and spacial and vertical symmetry.

2. History of nanosphere lithography

The journey of NSL started back in early 1983. Deckman and Dunsmuir [12] first published the fundamental concept of NSL which they called "natural lithography" by using monolayers of self-assembled ordered or random polystyrene spheres as a mask for patterning a substrate with nanoscale features. Later in 1995, Hulteen and Duyne [8] demonstrated the concept in detail for periodic arrays of particles creating triangular patterns with hexagonal symmetry connected or separated from each other through vertical material deposition. From that time the technique is famously known as "nanosphere lithography". In 1996, Winzer et al. [9] applied this technique to fabricate nano-dot and nano-ring arrays. The next year, periodic nanoholes with tunable diameter were demonstrated by Haginoya et al. [13] using reactive ion etching (RIE) assisted reduction of nanosphere mask followed by material deposition. Later in 1999, [14] size-tunable nanoparticles and nano-cluster arrays are created by varying the size of the nanosphere and the thickness of the deposited material. The concept of a bi-layer mask NSL is also introduced in this paper. In a venture to create ultra-flat nanostructures through NSL, Frey et al. [15] combined this technique with ultra-flat template stripping [16] in the year 2000. In this technique, a nanosphere mask is deposited on an ultra-flat mica substrate followed by material deposition. The nanosphere is removed leaving the bi-layer pattern on the mica similar to conventional NSL. In the next step, the pattern was covered by the deposition of a thin film adhesive layer and the subsequent application of epoxy on the top. Finally, mica is stripped off from underneath exposing an ultra-flat material pattern embedded in the matrix of the adhesive layer on the epoxy substrate. Further in the journey, NSL has evolved and experimented in association with other techniques. In 2004, more complex periodic structures are obtained using shadow NSL which is a union of shadow mask deposition and NSL [17]. In this study, structures are created by the controlled rotation of nanosphere shadow by tilting and rotating the substrate in the collimated shower of depositing material atoms. In 2005 arrays of rings, dots, and rods are demonstrated by shadow NSL with a deformed nanosphere mask obtained through annealing treatment [18]. In 2006, high aspect ratio nanopillars are created by NSL coupled with deep reactive ion etching (DRIE) [19]. The approach is based on the reduction of the size of the nanosphere mask followed by the DRIE of the substrate in the presence of the reduced mask. The technique is improvised and extended to create 3D hierarchical nanostructures like arrays of hollow cylinders, and step-like (solid or hollow) cylinders with decreasing diameter by performing multistep etching assisted NSL [20]. In 2007, Nano crescent [21] structures are created using nano template lithography (NTL) a simple extension of NSL. In this technique, instead of a large number of assembled nanospheres, individual nanospheres act as a template for angled deposition. The technique is further investigated for antenna application [22–24]. In 2015, a new variant of NSL called Moiré NSL (M-NSL) is developed which introduced in-plane rotation of neighboring monolayers creating meta-surfaces [25]. In the following years, the capability of M-NSL is expanded to create various superlattices with different materials aimed at diverse applications [26–29]. The concept of NSL is utilized to create ordered arrays of hollow nanospherical particle elements [30]. Self-assembled polystyrene nanospheres template is coated with fluorinated tin-oxide solution by slow drop casting. The liquid is dried, and the sample undergoes several steps of annealing to obtain the structure. One of the recent works on NSL finds a technique to pattern bridge networks between nanostructured arrays called nano-bridged NSL (NB-NSL) [31]. Further, step-wise formation of patterns using shadow mask deposition-based NSL is developed to precisely control the shape, size, gap between features, and overall symmetry of the complex 2D and 3D nanostructures [11, 32] using the azimuthal and polar angle of the shadow mask as a control parameter. As further extension of this work angular acceleration of a continuously rotating substrate has been introduced as functional control to the vertically asymmetric unique 3D structure [32]. Darvill et al. [10] described a scheme of etching-assisted NSL for structural symmetry breaking along the thickness.

The concept is based on anisotropic etching induced by the temperature gradient in the etching chamber.

The scope of NSL is not limited to the creation of a library of nanoarray structures. The knowledge base developed in this journey finds numerous opportunities for application-oriented research in various fields including plasmonics [4, 33–36], photonics [37], magnonics [38–42], optoelectronics [43], and biomedicine [44–48] which will be discussed in the application section.

3. Methods

In the last section, we have seen that the NSL technique is employed in various configurations in combination with several other techniques. In that sense, there is no unique recipe for NSL. However, the most standard form of NSL involves material deposition in the presence of a self-assembled nanosphere mask. Another form of NSL involves the etching of the nanosphere mask and the substrate. The process involves the following steps which are schematically presented in **Figure 1**.

In the first step, the substrate for instance Si(/SiO_2) wafer is cleaned using acetone and isopropyl alcohol (IPA) in an ultrasound bath, then washed with distilled water and dried by N_2 gas flow. Then the wafer undergoes Ar/O_2 plasma cleaning to eliminate organic impurities from the substrate and enhance its hydrophilicity. This treatment substantially affects the mobility of nanoparticle mask units, enabling them to distribute uniformly over the large surface of Si. In the second step, a solution of micro/nanobeads and ethanol is drop-casted on the substrate and allowed to be self-assembled in an HCP structure. Finally, the material is deposited usually by a sputterer or evaporator. Following the deposition, the sample undergoes ultrasonication in IPA to lift off the nanosphere masks and any material that has been deposited on them. The deposited material grains that can penetrate the gaps of the nanosphere assembly are left on the substrate as a lithographic structure as shown in the orange-shaded region of step 3 in **Figure 1**.

For etching-assisted NSL, the initial two steps remain unchanged. However, in the third step, the sample undergoes a two-step etching process instead of deposition as shown in the blue-shaded region of step 3 in **Figure 1**. Firstly, the nanospherical masks are treated with oxygen plasma in the RIE chamber. The polystyrene is etched by the oxygen plasma while the silicon substrate remains unaffected, causing the diameter of the beads to decrease. The diameter of the beads and their separation are determined by the etching duration. Subsequently, the exposed silicon is etched in the RIE/DRIE chamber while the silicon masked by the polystyrene beads remains unetched, resulting in a periodic nanostructured surface profile. Finally, the loosely bound polystyrene mask is removed from the substrate by immersing the sample in IPA and subjecting it to ultrasonic vibrations for 5 min. **Figure 1** displays a series of lithographed structures produced with NSL that are identified as 1–6 in the bottom row.

The hexagonal symmetry is an intrinsic property of the conventional which originates from the close packing of the nanospheres. When the nanospheres are deposited onto the substrate, they self-assemble into a HCP arrangement due to the minimization of the free energy of the system. This arrangement provides the highest packing density with the least amount of void space between the spheres. The self-assembly process is driven by the intermolecular force factors between the nanospheres and the interface, such as van der Waals force, Marangoni force (originating from the surface tension difference at the interface), zeta potential (ζ) electrostatic interaction forces. These forces cause the nanospheres to aggregate and form a well-defined

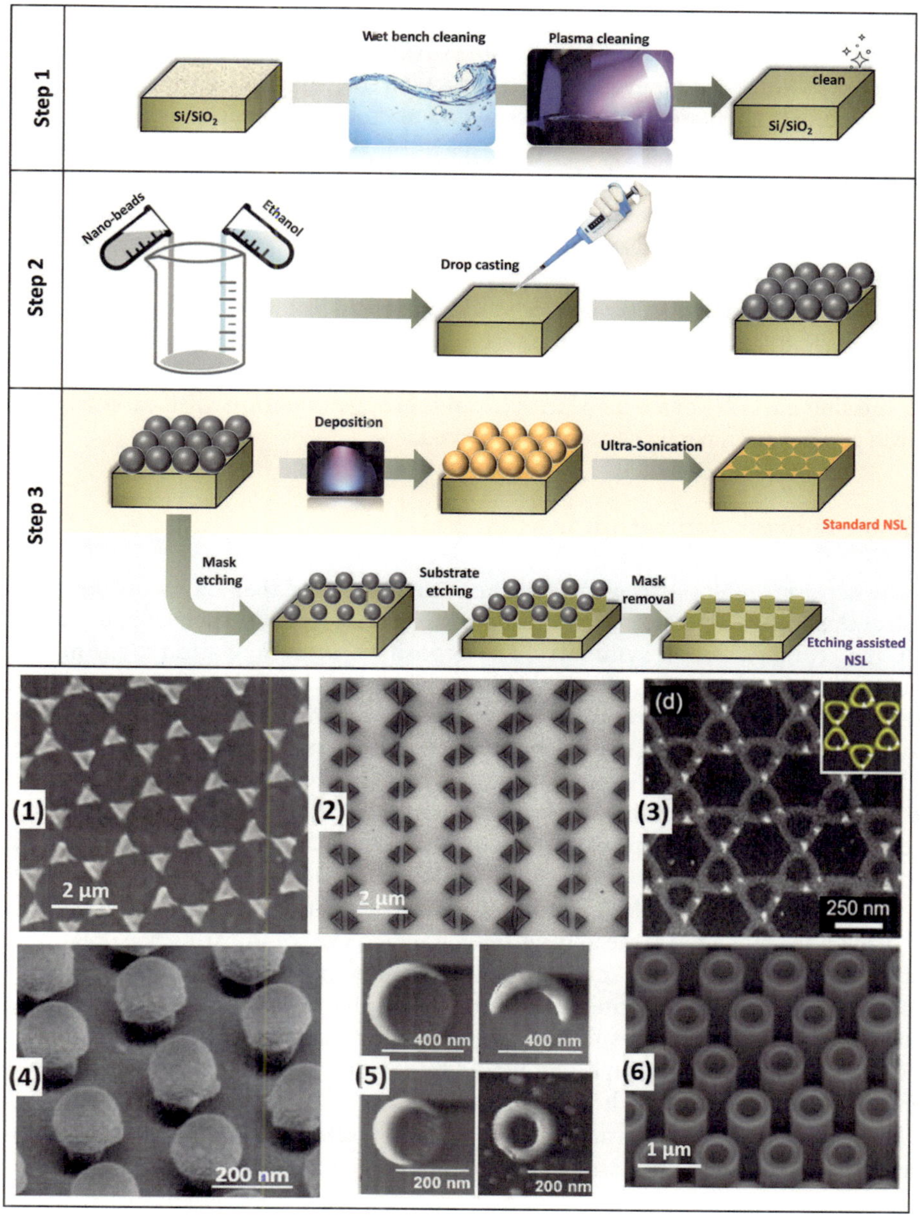

Figure 1.
Schematic diagram of the standard and etching-assisted NSL process along with an example of nanosphere lithographed structures (labeled as 1–6) at the bottom row [10, 11, 20, 21, 32]. (Ref 10—Creative Commons Attribution 3.0 Unported License, Ref 11—Creative Commons Attribution 4.0 license, Ref. 20, 21—Incorporated with permission and Ref 32—license under CCBY 4.0).

pattern on the substrate [49, 50]. The dynamics of self-assembly is a stochastic process having partial control over external parameters. Large area monolayer assembly of nanosphere can be achieved by optimizing a few control parameters. For instance, the temperature of the substrate can be optimized to control the evaporation rate of the solution drop-casted on it. Further, a rotating substrate with suitable angular speed applies centrifugal force on the nanospheres just sufficient for the close-pack monolayer formation. The self-assembly process is significantly influenced by the molar ratio of the solvent and the substrate's surface area. The solvent allows the micro/nanobeads to move until it evaporates, after which the particles become stationary. To ensure complete coverage of the surface with a monolayer particle mask, the volume of the drop-casted solution and the surface area of the substrate are meticulously selected to attain an appropriate particle density per unit square area. A higher concentration of micro/nanobeads leads to the formation of multilayers, while a lower density results in the creation of uncovered surface areas or islands. Apart from multilayer and island formation self-assembled nanospheres exhibit various lattice defects. Line defect is one of the most common defects which looks like a missing line in a matrix of nanospheres. Domain formation is another common form of defect where small domains of hexagonal lattices are formed with relative angle to each other. Apart from that, point defects (i.e., missing particles in a matrix array), random agglomeration, dislocations, and the uneven gap between particles are also observed.

4. Shadow-mask deposition assisted NSL

The deposition of a shadow mask is a powerful method that allows for the creation of a limitless range of structural patterns with distinct features such as size, shape, inter-structural gaps, and symmetry. The deposition angle in relation to the mask is a crucial factor in governing the direction and shape of the shadow, which ultimately determines the resulting structural pattern. In this technique, both azimuthal (θ) and polar (φ) angles are used as control parameters. The scanning electron microscope (SEM) image of a nanosphere mask arranged in HCP order is presented in **Figure 2a**, with the inset displaying the image at a higher magnification. **Figure 2b** explains the scheme of shadow mask deposition. The substrate on which the nanosphere is placed has a slope φ with respect to the horizontal plane. The red line indicates normal to the substrate and the vertical black line indicates the deposition vector. Hence, the red and black lines make the same angle φ between each other indicating the polar rotation of the substrate. **Figure 2c** illustrates the shadowing effect of a single nanosphere where green corresponds to the deposition of the substrate whereas red corresponds to the deposition of the nanosphere creating a shadow on the substrate.

The shadow structures can be simulated by various tools like Matlab or the light tracing software, Pov-Ray (The Persistence of Vision Ray-tracer). In this chapter, we use Matlab to simulate the structures. A program has been developed to solve equations related to straight lines and a group of spheres on a planar surface. The broken

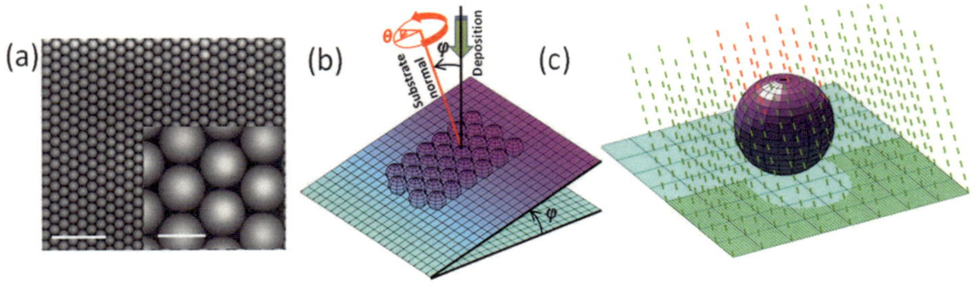

Figure 2.
(a) SEM image of nanosphere mask placed on a Si substrate, with the scale bars of the image and inset measuring 10 and 2 μm, respectively; (b) schematic diagram of the shadow mask deposition setup; and (c) Shadow of a single nanosphere [32]. (Ref. license under CCBY 4.0).

straight lines (**Figure 2c**) signify the shower of metal atoms with certain θ and φ, whereas the spheres symbolize the micro or nanobeads mask unit, and the plane corresponds to the substrate. When there is a real solution between a sphere and a straight line, it means that the evaporated beam is hitting the nanosphere (red), while an imaginary solution indicates that there is no contact between them (green). The first scenario corresponds to shadowing, while the latter implies that deposition on the substrate has occurred. In the latter case, the straight line and substrate plane are further solved to determine the coordinates of the deposition.

4.1 Single-step NSL

Figure 3a shows a matrix of nanostructures with systematically varying θ and φ. The largest pattern is observed for the column φ = 0 having surface exposure of 7.59% of the substrate area. The patterns become smaller with increasing φ as we move from the left to the right of the row for each value of θ. For instance, at φ = 40 degrees and θ = 0 degree the exposure reduces to 0.62%. When the value of φ is increased, the deposition vector becomes more slanted in relation to the substrate. As a result, a greater portion of the substrate is shaded by the nanosphere, resulting in a smaller pattern size. The curves in **Figure 3b** show the variation in the exposure % as a function of φ for θ = 0 and 90 degrees. The latter (θ = 90 degrees) has a larger exposure over the range of φ. The theoretical value of the minimum tilt angle for which the nanosphere completely covers the substrate is defined as the extinction angle (φ). For the two perpendicular θ (=0 and 90 degrees) φ_e is calculated as 46 and 51 degrees respectively. The importance of this parameter is that for shadow mask deposition-assisted NSL the magnitude of φ must be lower than φ_e. The pattern is hexagonal symmetric for φ = 0 (i.e., normal incidence) as shown by the white lines in **Figure 3**. The pattern remains exactly the same for any angle θ meaning the configuration is azimuthally symmetric. The symmetry breaks for a nonzero value of φ. The elements are shifted such that some neighboring structures get closer, and some get farther from each other giving rise to a polarized pattern as indicated by the red arrows in **Figure 3**.

4.2 NSL with varying φ

Here we show 3D nanostructures with step-like shapes that are achieved by varying the angle φ while keeping θ constant, as illustrated in panels 1 and 2 of **Figure 4**. The

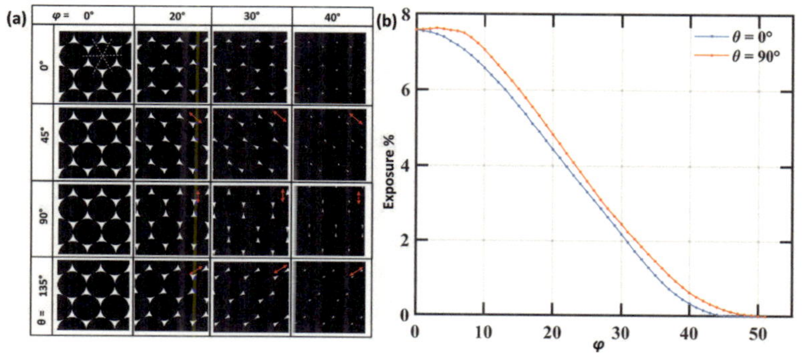

Figure 3.
(a) Simulated structure matrix with systematically varying θ and φ. (b) Exposured area (%) with varying φ at θ = 0 degree and θ = 90 degrees.

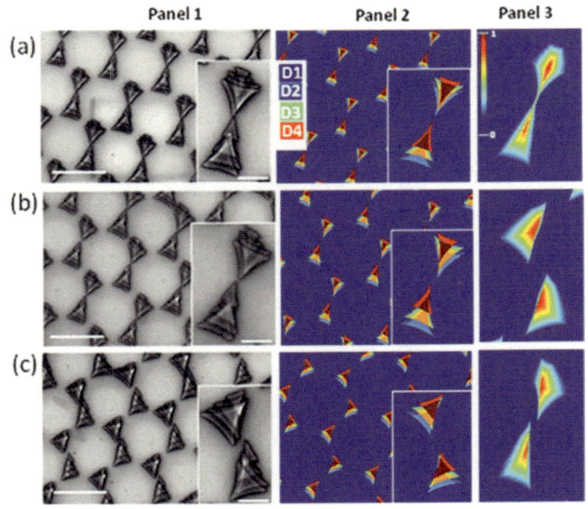

Figure 4.
The deposition of shadows in four steps, with different values of φ (10, 15, 20, and 25 degrees), keeping θ constant at (a) 20 degrees, (b) 0 degree and (c) 45 degrees. The scale bars in the figure and in the insets in panel 1 correspond to 2 μm and 500 nm, respectively. Panel 2 and 3 show the simulation result for four steps of φ angle and continuously varying φ angle deposition [32]. (Ref. license under CCBY 4.0).

inset in each figure shows a magnified image of the lattice. Panel 1 of **Figure 4** displays SEM images of the four steps of deposition, in which φ is varied as 10, 15, 20, and 25 degrees at constant θ (**Figure 4a** 20 degrees, **4b** 0 degree and **4c** 45 degrees). The resulting structure consists of four curved arm triangles of different sizes and aspect ratios, which are superposed with a gradual shift. Each deposition step serves as a building block to create a vertically asymmetric 3D nanostructure. Panel 2 of **Figure 4** presents the simulated result, which agrees with the experimental observation. Different colors are used to indicate the four depositions (D1–D4) in this figure. In contrast, when the deposition is carried out by continuously varying φ, rather than in steps, unique 3D structures with smoothly varying surface topology are formed, as shown in panel 3 of **Figure 4**. These structures were simulated for constant θ, with φ varying from 0 to 30 degrees in 1 degree steps at a constant angular velocity. A further variation of shape can be obtained by introducing angular acceleration to the rotating substrate.

4.3 Two-step NSL with varying θ

Lattice structures obtained from two-fold azimuthal symmetry are shown in **Figure 5**. Unlike the previous case, this configuration performs a two-step deposition with φ as constant and θ varying as θ_i and θ_i + 180 degrees, where θ_i stands for the θ of the first deposition. Panel 1 and 2 correspond to the experimental and simulated results which are consistent with each other. The patterns break the conventional hexagonal symmetry with the rearrangement of elements rather explained by linear symmetry. The values of θ and φ are mentioned for each figure. **Figure 5a** shows pairs of triangles arranged in a zigzag pattern around regularly spaced line gaps. In **Figure 5b** and **c**, the unit cell structures are made up of four triangles combined. **Figure 5d** displays square structures that are intersected by line gaps. **Figure 5e** exhibits a mixture of square and triangle shapes, while **Figure 5f** is composed of a combination of star and triangular arrays.

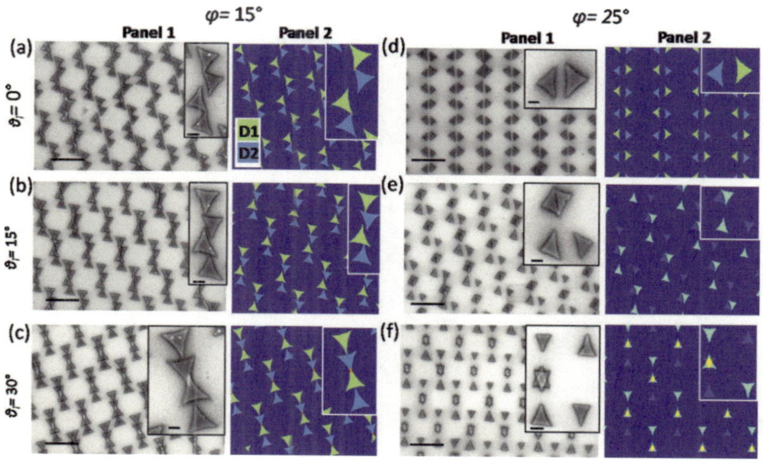

Figure 5.
Two-step deposition for varying θ (θi and θi + 180 degrees) at constant φ. The φ value for figure (a–c) is 15 degrees, and for figure (d–f) is 25 degrees. Panels 1 and 2 show the SEM and simulation results respectively. In panel 1, the scale bars displayed in the figures and insets are 2 μm and 500 nm, respectively [32]. (Ref. license under CCBY 4.0).

4.4 Complex structures with varying θ and φ

The possibilities for generating diverse lattices through shadow mask deposition-aided NSL are practically endless. In this section, examples of some complex lattice structures are shown with uncommon shapes and symmetry. In **Figure 6a** and **b**, the nanostructures are produced with four-fold symmetry of θ. It is a quantitative addition of two times of two-fold symmetry perpendicular to each other configured as θ = θi + (0, 90, 180, and 270 degrees). **Figure 6a** and **b** are the SEM images for two different θi (15 and 355 degrees respectively) at a constant φ of 25 degrees. A bunch of eight triangles reorient and restructure themselves to form a unit cell as shown in the insets. Further, the structure in **Figure 6c** is obtained with φ = 25 degrees and θ = 0 and 90 degrees. The deposition is neither two-fold nor four-fold symmetric to θ rather can be termed as incomplete (or semi) four-fold symmetric. **Figure 6d** is obtained by a four-step deposition of simultaneously varying θ and φ as θ = 130 degrees + (0, 90, 180,

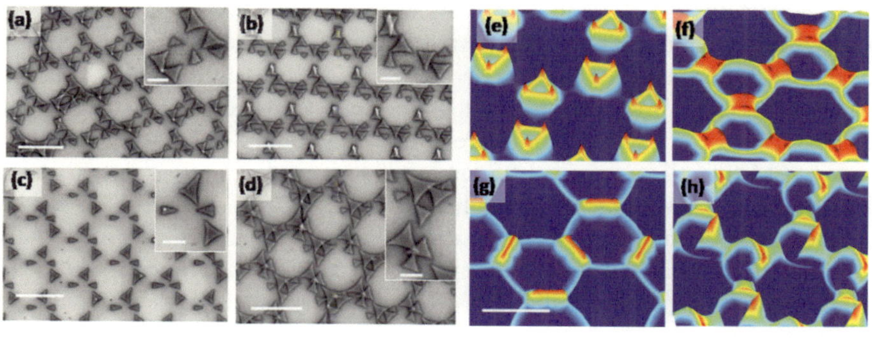

Figure 6.
(a–d) SEM images of complex lattice structures obtained by multistep deposition using various combinations of θ and φ. The scale bars in the figure and the insets correspond to 2 μm and 500 nm, respectively. Simulated structures upon complete rotation of θ for (e-g) constant φ and (h) continuously varying φ [32]. (Ref. license under CCBY 4.0).

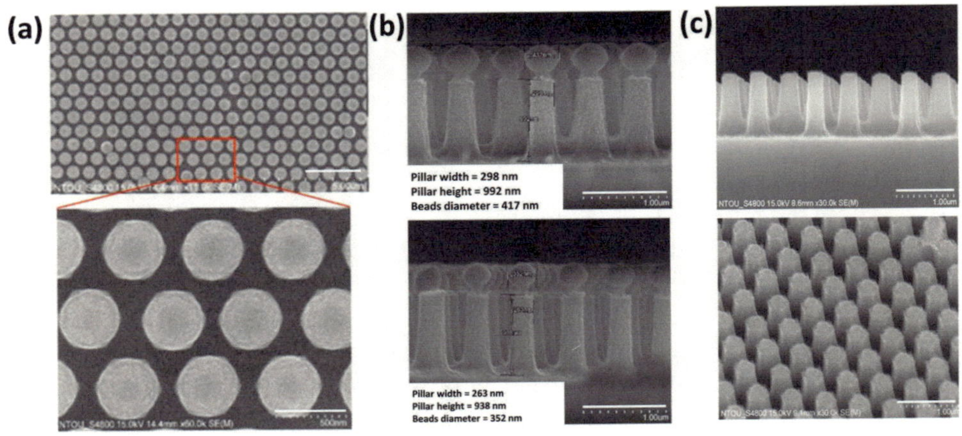

Figure 7.
SEM images of the sample at different stages of etching-assisted NSL. (a) Top: nanospheres on Si substrate after oxygen etching. Bottom: Magnified image of the red square marked region. White lines in the top and bottom images correspond to 2 μm and 500 nm respectively. (b) Side view of nanopillars array with etched nanospheres on the top of each pillar. The top and bottom images correspond to 4 and 5 min of oxygen etching respectively. (c) Side and slanted view images of nanopillars array after removal of nanosphere mask. Scale bars in figures (b) and (c) correspond to 1 μm [4]. (Ref. license under CCBY 4.0).

and 270 degrees) and φ = 15, 18, 22and 29 degrees. In **Figure 6e–h**, simulated structures are obtained by integrating over the full rotation of θ. The simulation involves varying θ from 0 to 359 degrees in increments of 1 degree. In **Figure 6e–h**, the value of φ is kept constant at 15, 25, and 30 degrees respectively, while in **Figure 6h**, φ is continuously increasing from 15 to 30 degrees at a constant angular velocity over the entire span of θ. **Figure 6e-h** shows a series of unique 3D patterns that are interconnected in a hexagonal framework. Among these patterns, the nano spiral shape illustrated in **Figure 6h** is particularly important and could be employed in chiral spectroscopy.

5. Etching-assisted NSL

Figure 7 presents the summary of the SEM results obtained from etching-assisted NSL [4]. The image at the top of **Figure 7a** displays a self-assembled nanosphere mask following size reduction. Etching of the nanosphere mask is achieved using oxygen plasma in the RIE chamber. This process reduces the diameter of the nanospheres and evenly spaces them apart. The spacing between nanospheres increases as the size of the nanospheres reduces with a longer exposure time to the oxygen plasma. The image at the bottom offers a closer view of the region marked by the red rectangle. **Figure 7b** showcases SEM side view images of a Si nanopillar array after undergoing Si etching in the presence of a size-reduced nanosphere mask. In this process, the exposed Si is etched from the top while the Si underneath the nanosphere beads remains protected, resulting in a cylindrical nanopillar array pattern. The process involves inductively coupled plasma reactive ion etching (ICP-RIE). Alternatively, the time-multiplexed silicon etching, commonly referred to as "Bosch" can be employed in a Deep RIE chamber to achieve high aspect ratio pillars with minimal etching of the side walls [19]. In this technique, Si is etched with SF_6, followed by passivation with a C_4F_8 polymer in repeated cycles. The top and bottom images of **Figure 7b** correspond to different time exposure to oxygen plasma (4 and 5 min, respectively). The spherical particles seen on top of each pillar represent the nanosphere masks. Ideally, the diameter of the cylindrical pillars should be the same as the size of the nanosphere. However, in this case, it appears slightly smaller due to nonuniformity in the etching process. Comparing the top and bottom figures, it is evident that the pillars are narrower in the bottom image due to reduced beads size. Thus, the size of the nanopillars can be controlled by the etching time. Finally, the top and bottom images of **Figure 7c** present side and slanted view images of the pillars after the removal of the nanosphere beads.

The conventional etching-assisted NSL technique has been improved to create a hollow cylindrical pillar array through multiple-patterning nanosphere lithography (MP-NSL) [20], as shown in **Figure 8a**. The first three steps of MP-NSL are the same as the standard etching-assisted NSL, where nanospheres on a Si substrate are treated with oxygen plasma to reduce their diameter, followed by DRIE to create solid cylindrical arrays. The diameter of the reduced nanosphere will be the same as the outer diameter of the resulting hollow cylinder. In step 4, the nanospheres are treated with oxygen plasma again, further reducing their diameter, which will become the inner diameter of the cylinder. The reduction of diameter in this step determines the thickness of the hollow cylinder. Next, Ni is deposited from the top, and the nanosphere is removed, leaving a ring-shaped mask at the top of the solid cylinder along its outer diameter (steps 4 and 5). Finally, in steps 6 and 7, the inner hole is created through DRIE, and the Ni ring is removed to obtain a hollow cylindrical array of Si. In **Figure 8b** SEM images labeled 1–6 are shown corresponding to the lithography steps. Label 1 shows closely packed monolayer nanospheres of diameter 1 μm on Si substrate (top view). In label 2 nanospheres are shown with the first reduction of diameter (top view). In label 3, a tilted view (tilt angle = 30 degrees) of solid pillars of Si with nanospheres on top is shown. Label 4 shows the size of the nanosphere on top

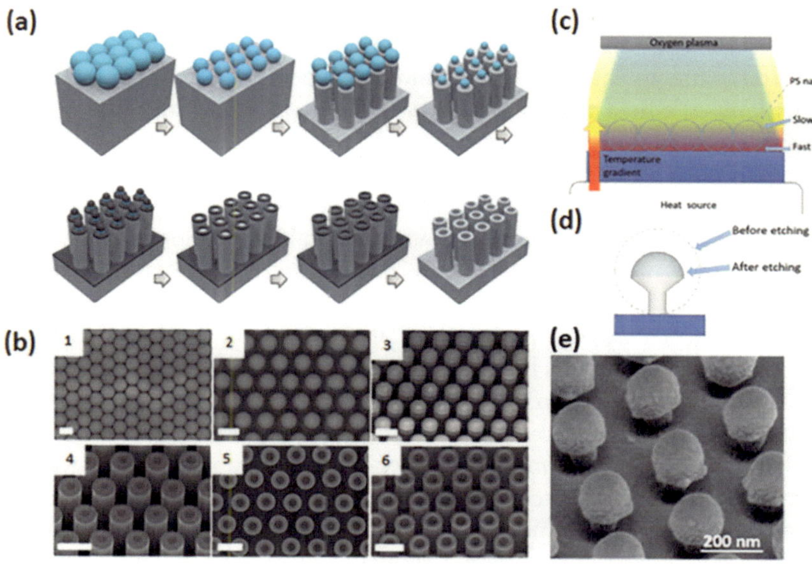

Figure 8.
(a) Schematic illustration of the MP-NSL process. (b) 1. Nanosphere mask on a Si wafer; 2. nanospheres after the first size reduction; 3. nanospheres on top of Si nanopillar arrays; 4. second size reduction of nanospheres; 5. Ni nanorings on top of nanopillars and 6. etching of the inner regions by DRIE forming Si nanotubes. Scale bars are 1 μm [20]. (c) Scheme of vertical temperature gradient induced anisotropic etching. (d) Morphology of nano mushroom. (e) SEM image of gold-coated nano mushroom arrays [10]. (Ref. 10—Creative Commons Attribution 3.0 Unported License, Ref. 20—Incorporated with permission).

of Si pillars is reduced further (tilted view). The top view after Ni coating is shown in label 5. Finally, in label 6 tilted views of hollow nanosphere are shown. In the tilted view images, the tilt angle is kept at 30 degrees. The scale bars correspond to 1 μm.

In the previous discussion we have seen etching to be isotropic, however, controlled anisotropy can be introduced in the etching process by creating a temperature gradient inside the etching chamber, resulting in broken symmetry along the thickness of a structure. The process of anisotropic plasma etching involves an isotropic oxygen plasma etching along with a vertical temperature gradient that is applied from the bottom to the top of the nanospheres. The rate of etching is dependent on the chamber temperature, with higher temperatures resulting in a faster rate of etching. A uniform change in chamber temperature does not change the shape of the structure, only the speed of the etching process. By introducing the temperature gradient, the etching rate can be engineered along the thickness, resulting in a highly anisotropic etching process perpendicular to the substrate [10]. **Figure 8c** depicts the setup of the etching chamber, while **Figure 8c** and **d** schematically illustrate how a nanosphere transforms into a mushroom-shaped structure, with the lower hemisphere which is in contact with the substrate eroding faster due to heat accumulation from the higher temperature. Finally, **Figure 8e** shows a slanted view SEM image of the resulting nano mushroom array coated with a thin film of gold.

6. Application

From the discussion above we found NSL to be a powerful tool to create ordered arrays of nanoscale features with controlled size, shape, and spacing. Nanostructures having nanodomains that extend down to a few nm are crucial for plasmonics, particularly in applications such as surface-enhanced Raman spectroscopy (SERS), chiral spectroscopy, and biosensors. The applications rely on surface plasmons creating anisotropic optical behavior to manipulate light at the nanoscale. These structures from NSL can concentrate the electric field at an extremely small apex (radius of curvature < 10 nm) or vertices, resulting in amplified optical signals. It is widely used to create SERS substrates, which can enhance the Raman scattering of molecules from ultra-low concentration with high sensitivity and specificity. SERS has applications in the detection of molecules, proteins, and cells for chemical detection, environmental monitoring, and medical diagnosis. For example, a recent report [51] in 2020, demonstrates early detection of liver cancer (hepatocellular carcinoma) using the SERS technique. The study uses a special bridged knobby nanostructured template fabricated by NSL for the detection of low-concentration cancer bio-markers [52]. NSL patterning can be used for high-resolution imaging of a live cell or a single molecule using the localized surface plasmon resonance (LSPR) technique [53]. The research conducted centers around distinct metallic nanostructures that possess varying symmetries, inter-structural gaps, and chirality. These structures are studied in order to regulate electromagnetic hotspots and adjust optical characteristics based on the alignment of their constituent elements [54–59]. Farther more, the effectiveness of magnetic memory applications such as data storage, data transfer, and logic devices are reliant on the regulated transmission of spin waves through magnonic crystals via exchange interactions. Over the last decade, there has been considerable research conducted in controlling the chiral magnetic domains [60, 61] as well as spin wave properties like frequency modes and damping [62, 63] by modifying the thickness and shape anisotropy of nanostructures [64, 65]. Hence, nanostructures, rich in various spin-wave modes, are of high importance [66–70]. NSL has also been used to fabricate plasmonic color filters [71, 72], waveguides [73, 74], and antennas [74–76] with high efficiency and tunability [71]. Moreover, smart window application employs a lithographed pattern of thermochromic material by NSL to control the transmittance of the window material [77]. NSL can be used to create periodic texture on the surface of solar cells, which can increase light absorption and thereby improve the efficiency of the solar cells [78, 79]. It can also be used for fabricating antibacterial surfaces [80]. Arrays of nanoneedles fabricated by NSL are demonstrated for intracellular delivery of nanoparticles with an attached microfluidic system for cellular immunotherapies [81].

7. Conclusion

To summarize, NSL is a micro/nanofabrication technique that is inexpensive and straightforward, offering numerous benefits over other lithography techniques such as scalability, versatility, time efficiency, and eco-friendliness. Consequently, it has become a widely popular technique, with substantial research efforts invested over the last few decades. As the technique has developed over time, it has progressed from simple hexagonal patterns of triangular elements to more complex 2D/3D structure fabrication with controlled shape, symmetry, and inter-structural spacing. To create various nanostructures for different applications, the NSL method has been modified and associated with various other techniques. The NSL method can be broadly categorized into two parts: deposition-assisted NSL and etching-assisted NSL. The former involves depositing material on closely packed nanosphere masks to create a shadow pattern on the substrate, while the latter involves patterning the substrate by etching it through the gaps of the nanosphere. NSL can also be performed with a complex rotation (either continuous or discrete) of the substrate, producing unique patterns such as star and spiral patterns. Additionally, a vertical temperature gradient during etching induces symmetry breaking along the thickness, which is required for engineering 3D structures. The NSL technique finds significant applications in the field of plasmonics, magnonics, and biomedicine.

Author details

Arnab Ganguly[1] and Gobind Das[1,2*]

1 Department of Physics, Khalifa University, Abu Dhabi, United Arab Emirates

2 Center for Catalysis and Separation (CeCas), Khalifa University of Science and Technology, Abu Dhabi, United Arab Emirates

*Address all correspondence to: gobind.das@ku.ac.ae

References

[1] De Angelis F, Das G, Candeloro P, Patrini M, Galli M, Bek A, et al. Nanoscale chemical mapping using three-dimensional adiabatic compression of surface plasmon polaritons. Nature Nanotechnology. 2010;**5**:67-72

[2] Das G, Battista E, Manzo G, Causa F, Netti PA, Di Fabrizio E. Large-scale plasmonic nanocones array for spectroscopy detection. ACS Applied Materials & Interfaces. 2015;**7**:23597-23604

[3] Das G, Patra N, Gopalakrishnan A, Zaccaria RP, Toma A, Thorat S, et al. Fabrication of large-area ordered and reproducible nanostructures for SERS biosensor application. The Analyst. 2012;**137**:1785-1792

[4] Chou Chau Y-F, Chen K-H, Chiang H-P, Lim CM, Huang HJ, Lai C-H, et al. Fabrication and characterization of a metallic–dielectric nanorod array by nanosphere lithography for plasmonic sensing application. Nanomaterials. 2019;**9**:1691

[5] Waegner M. Nanosphere Lithography. In: Gerlach G, Wolter KJ, editors. Bio and Nano Packaging Techniques for Electron Devices: Advances in Electronic Device Packaging. Berlin, Heidelberg: Springer; 2012. p. 269-277. DOI: 10.1007/978-3-642-28522-6_12

[6] Li J, Hu Y, Yu L, Li L, Ji D, Li L, et al. Recent advances of nanospheres lithography in organic electronics. Small. 2021;**17**:2100724

[7] Fang X, Zheng C, Yin Z, Wang Z, Wang J, Liu J, et al. Hierarchically ordered silicon metastructures from improved self-assembly-based nanosphere lithography. ACS Applied Materials & Interfaces. 2020;**12**:12345-12352

[8] Hulteen JC, Van Duyne RP. Nanosphere lithography: a materials general fabrication process for periodic particle array surfaces. Journal of Vacuum Science and Technology A. 1995;**13**:1553-1558

[9] Winzer M, Kleiber M, Dix N, Wiesendanger R. Fabrication of nano-dot-and nano-ring-arrays by nanosphere lithography. Applied Physics A: Materials Science & Processing. 1996;**63**:617-619

[10] Darvill D, Iarossi M, Ekeroth RMA, Hubarevich A, Huang J-A, De Angelis F. Breaking the symmetry of nanosphere lithography with anisotropic plasma etching induced by temperature gradients. Nanoscale Advances. 2021;**3**:359-369

[11] Myint B, Yap DSF, Ng V. Stepwise nanosphere lithography: an alternate way of fabricating nanostructures. Nano Express. 2020;**1**:020029

[12] Deckman H, Dunsmuir J. Natural lithography. Applied Physics Letters. 1982;**41**:377-379

[13] Haginoya C, Ishibashi M, Koike K. Nanostructure array fabrication with a size-controllable natural lithography. Applied Physics Letters. 1997;**71**:2934-2936

[14] Hulteen JC, Treichel DA, Smith MT, Duval ML, Jensen TR, Van Duyne RP. Nanosphere lithography: size-tunable silver nanoparticle and surface cluster arrays. The Journal of Physical Chemistry. B. 1999;**103**:3854-3863

[15] Frey W, Woods CK, Chilkoti A. Ultraflat nanosphere lithography: a new

method to fabricate flat nanostructures. Advanced Materials. 2000;**12**:1515-1519

[16] Wagner P, Hegner M, Guentherodt H-J, Semenza G. Formation and in situ modification of monolayers chemisorbed on ultraflat template-stripped gold surfaces. Langmuir. 1995;**11**:3867-3875

[17] Kosiorek A, Kandulski W, Chudzinski P, Kempa K, Giersig M. Shadow nanosphere lithography: simulation and experiment. Nano Letters. 2004;**4**:1359-1363

[18] Kosiorek A, Kandulski W, Glaczynska H, Giersig M. Fabrication of nanoscale rings, dots, and rods by combining shadow nanosphere lithography and annealed polystyrene nanosphere masks. Small. 2005;**1**:439-444

[19] Cheung CL, Nikolić R, Reinhardt C, Wang T. Fabrication of nanopillars by nanosphere lithography. Nanotechnology. 2006;**17**:1339

[20] Xu X, Yang Q, Wattanatorn N, Zhao C, Chiang N, Jonas SJ, et al. Multiple-patterning nanosphere lithography for fabricating periodic three-dimensional hierarchical nanostructures. ACS Nano. 2017;**11**:10384-10391

[21] Bukasov R, Shumaker-Parry JS. Highly tunable infrared extinction properties of gold nanocrescents. Nano Letters. 2007;**7**:1113-1118

[22] Bukasov R, Ali TA, Nordlander P, Shumaker-Parry JS. Probing the plasmonic near-field of gold nanocrescent antennas. ACS Nano. 2010;**4**:6639-6650

[23] Rodriguez M, Furse C, Shumaker-Parry JS, Blair S. Scaling the response of nanocrescent antennas into the ultraviolet. ACS Photonics. 2014;**1**:496-506

[24] Swartz M, Rodriguez M, Quast AD, Cooper CT, Blair S, Shumaker-Parry JS. Aluminum nanocrescent plasmonic antennas fabricated by copper mask nanosphere template lithography. Journal of Physical Chemistry C. 2016;**120**:20597-20603

[25] Chen K, Rajeeva BB, Wu Z, Rukavina M, Dao TD, Ishii S, et al. Moiré nanosphere lithography. ACS Nano. 2015;**9**:6031-6040

[26] Wu Z, Li W, Yogeesh MN, Jung S, Lee AL, McNicholas K, et al. Tunable graphene metasurfaces with gradient features by self-assembly-based Moiré nanosphere lithography. Advanced Optical Materials. 2016;**4**: 2035-2043

[27] Chen K, Rajeeva BB, Wu Z, Rukavina M, Dao TD, Ishii S, Aono M, et al. Moiré nanosphere lithography: use colloidal moiré patterns as masks. In: Plasmonics: Metallic Nanostructures and Their Optical Properties XIII, SPIE. Vol. 9547. 2015. pp. 209-214

[28] Wu Z, Kelp G, Yogeesh MN, Li W, McNicholas KM, Briggs A, et al. Dual-band moiré metasurface patches for multifunctional biomedical applications. Nanoscale. 2016;**8**:18461-18468

[29] Wu Z, Zheng Y. Moiré metamaterials and metasurfaces. Advanced Optical Materials. 2018;**6**:1701057

[30] Robbiano V, Paternò G, Cotella G, Fiore T, Dianetti M, Scopelliti M, et al. Polystyrene nanoparticle-templated hollow titania nanosphere monolayers as ordered scaffolds. Journal of Materials Chemistry C. 2018;**6**:2502-2508

[31] Luo L, Akinoglu EM, Wu L, Dodge T, Wang X, Zhou G, et al. Nano-bridged nanosphere lithography. Nanotechnology. 2020;**31**:245302

[32] Ganguly A, Das G. Combining azimuthal and polar angle resolved shadow mask deposition and nanosphere lithography to uncover unique nano-crystals. Nanomaterials. 2022;**12**:3464

[33] Hicks EM, Zhang X, Zou S, Lyandres O, Spears KG, Schatz GC, et al. Plasmonic properties of film over nanowell surfaces fabricated by nanosphere lithography. The Journal of Physical Chemistry. B. 2005;**109**:22351-22358

[34] Jensen TR, Malinsky MD, Haynes CL, Van Duyne RP. Nanosphere lithography: tunable localized surface plasmon resonance spectra of silver nanoparticles. The Journal of Physical Chemistry. B. 2000;**104**:10549-10556

[35] Rahaman M, Moras S, He L, Madeira TI, Zahn DR. Fine-tuning of localized surface plasmon resonance of metal nanostructures from near-Infrared to blue prepared by nanosphere lithography. Journal of Applied Physics. 2020;**128**:233104

[36] Lu Y-C, Hsueh C-H. Plasmonic and optical properties of periodic silver nanoprism array fabricated by H_2O_2-assisted nanosphere lithography. In: JSAP-OSA Joint Symposia, Optical Society of America, 2018, p. 18p_211B_213

[37] Li Z, Song C, Li Q, Xiang X, Yang H, Wang X, et al. Hybrid nanostructured antireflection coating by self-assembled nanosphere lithography. Coatings. 2019;**9**:453

[38] Tiberto P, Barrera G, Celegato F, Conta G, Coisson M, Vinai F, et al. Ni80Fe20 nanodisks by nanosphere lithography for biomedical applications. Journal of Applied Physics. 2015;**117**:17B304

[39] Semenova E, Montoncello F, Tacchi S, Dürr G, Sirotkin E, Ahmad E, et al. Magnetodynamical response of large-area close-packed arrays of circular dots fabricated by nanosphere lithography. Physical Review B. 2013;**87**:174432

[40] Montoncello F, Tacchi S, Giovannini L, Madami M, Gubbiotti G, Carlotti G, et al. Asymmetry of spin wave dispersions in a hexagonal magnonic crystal. Applied Physics Letters. 2013;**102**:202411

[41] Manzin A, Barrera G, Celegato F, Coïsson M, Tiberto P. Influence of lattice defects on the ferromagnetic resonance behaviour of 2D magnonic crystals. Scientific Reports. 2016;**6**:1-11

[42] Krupiński M, Zabila Y, Marszalek M. Nanopatterned Thin Films with Perpendicular Magnetic Anisotropy–Structure and Magnetism. In: Kaidatzis A, Sidorenko S, Vladymyrskyi I, Niarchos D, editors. Modern Magnetic and Spintronic Materials. Netherlands: Springer; 2020. p. 47-71. DOI: 10.1007/978-94-024-2034-0_3

[43] Qiu T, Akinoglu EM, Luo B, Konarova M, Yun JH, Gentle IR, et al. Nanosphere lithography: a versatile approach to develop transparent conductive films for optoelectronic applications. Advanced Materials. 2022;**34**:2103842

[44] Tian F, Li M, Wu S, Li L, Hu H. A hybrid and scalable nanofabrication approach for bio-inspired bactericidal silicon nanospike surfaces. Colloids and Surfaces. B, Biointerfaces. 2023;**222**:113092

[45] Lisboa P, Villiers M-B, Brakha C, Marche PN, Valsesia A, Colpo P, et al. Fabrication of bio-functionalised polypyrrole nanoarrays for bio-molecular recognition. Micro and Nanosystems. 2011;**3**:83-89

[46] Cai Y, Ocko BM. Large-scale fabrication of protein nanoarrays based on nanosphere lithography. Langmuir. 2005;**21**:9274-9279

[47] Xiang G, Zhang N, Zhou X. Localized surface plasmon resonance biosensing with large area of gold nanoholes fabricated by nanosphere lithography. Nanoscale Research Letters. 2010;**5**:818-822

[48] Purwidyantri A, Chen C-H, Hwang B-J, Luo J-D, Chiou C-C, Tian Y-C, et al. Spin-coated Au-nanohole arrays engineered by nanosphere lithography for a Staphylococcus aureus 16S rRNA electrochemical sensor. Biosensors & Bioelectronics. 2016;**77**:1086-1094

[49] Cossio G, Yu ET. Zeta potential dependent self-assembly for very large area nanosphere lithography. Nano Letters. 2020;**20**:5090-5096

[50] Scriven L, Sternling C. The marangoni effects. Nature. 1960;**187**:186-188

[51] Cheng M, Zhang F, Zhu A, Zhang X, Wang Y, Zhao X, et al. Bridging the neighbor plasma coupling on curved surface array for early hepatocellular carcinoma detection. Sensors & Actuators, B: Chemical. 2020;**309**:127759

[52] Hudson SD, Chumanov G. Bioanalytical applications of SERS (surface-enhanced Raman spectroscopy). Analytical and Bioanalytical Chemistry. 2009;**394**:679-686

[53] Lee S. Enhanced, Live-Cell, Super-Resolution Imaging and Single-Molecule Emission Spectrum Reshaping [thesis]. The University of Michigan; 2019

[54] Zhao Y, Chi Y, Hong Y, Li Y, Yang S, Yin J. Twisting for soft intelligent autonomous robot in unstructured environments. Proceedings of the National Academy of Sciences. 2022;**119**:e2200265119

[55] Dang A, Wang Y, Zhang H, Panatdasirisuk W, Xia Y, Wang Z, et al. Electrospun gold nanoprism/poly(vinyl alcohol) nanofibers for flexible and free-standing surface-enhanced raman scattering substrates. ACS Applied Nano Materials. 2022;**5**:6650-6658

[56] Chirumamilla M, Toma A, Gopalakrishnan A, Das G, Zaccaria RP, Krahne R, et al. 3D nanostar dimers with a sub-10-nm gap for single−/few-molecule surface-enhanced Raman scattering. Advanced Materials. 2014;**26**:2353-2358

[57] Langer J, Jimenez de Aberasturi D, Aizpurua J, Alvarez-Puebla RA, Auguié B, Baumberg JJ, et al. Present and future of surface-enhanced Raman scattering. ACS Nano. 2019;**14**:28-117

[58] Cecchini MP, Turek VA, Paget J, Kornyshev AA, Edel JB. Self-assembled nanoparticle arrays for multiphase trace analyte detection. Nature Materials. 2013;**12**:165-171

[59] Das G, Mecarini F, Gentile F, De Angelis F, Mohan Kumar HG, Candeloro P, et al. Nano-patterned SERS substrate: application for protein analysis vs. temperature. Biosensors & Bioelectronics. 2009;**24**:1693-1699

[60] Ganguly A, Zhang S, Miron IM, Kosel J, Zhang X, Manchon A, et al. Das G.: Competition between chiral energy and chiral damping in the asymmetric expansion of magnetic bubbles. ACS Applied Electronic Materials. 2021;**3**:4734-4742

[61] Zhang S, Zhang X, Zhang J, Ganguly A, Xia J, Wen Y, et al. Direct

imaging of an inhomogeneous electric current distribution using the trajectory of magnetic half-skyrmions. Science Advances. 2020;**6**:eaay1876

[62] King J, Ganguly A, Burn D, Pal S, Sallabank E, Hase T, et al. Local control of magnetic damping in ferromagnetic/non-magnetic bilayers by interfacial intermixing induced by focused ion-beam irradiation. Applied Physics Letters. 2014;**104**:242410

[63] Ganguly A, Azzawi S, Saha S, King J, Rowan-Robinson R, Hindmarch A, et al. Tunable magnetization dynamics in interfacially modified Ni81Fe19/Pt bilayer thin film microstructures. Scientific Reports. 2015;**5**:1-8

[64] Rana B, Ganguly A, Barman A. Magnetic shape anisotropy in chemically synthesized chains of nickel nanoparticles. IEEE Transactions on Magnetics. 2011;**47**:2859-2862

[65] Mahato B, Ganguly A, Rana B, Barman A. Magnetization reversal in chemically synthesized hexagonal cobalt microplatelets. Journal of Physical Chemistry C. 2012;**116**:22057-22062

[66] Imperia P, Kandulski W, Kosiorek A, Głaczyńska H, Maletta H, Giersig M. Magnetic anisotropy study of triangular-shaped Co nanostructures. Journal of Magnetism and Magnetic Materials. 2008;**320**:2682-2687

[67] Donnelly C, Finizio S, Gliga S, Holler M, Hrabec A, Odstrčil M, et al. Time-resolved imaging of three-dimensional nanoscale magnetization dynamics. Nature Nanotechnology. 2020;**15**:356-360

[68] Caso D, Aliev FG. Edge spin wave transmission through a vertex domain wall in triangular dots. SN Applied Sciences. 2022;**4**:1-9

[69] da Silva J, Guerra Y, Padrón-Hernández E. Simulations of FMR for study the shape anisotropy in square hollow nanopillars. Journal of Magnetism and Magnetic Materials. 2022;**544**:168642

[70] Jin C, Li S, Zhang H, Wang R, Wang J, Lian R, et al. Spin-wave modes of elliptical skyrmions in magnetic nanodots. New Journal of Physics. 2022;**24**:043005

[71] Liu J, Cai H, Tang K, Kong L. Plasmonic filters based on lithographically patterned and hexagonally arranged triangular silver nanoparticles array. Advances in Materials Science and Engineering. 2015;**2015**:195320

[72] Proietti Zaccaria R, Bisio F, Das G, Maidecchi G, Caminale M, Vu CD, et al. Plasmonic color-graded nanosystems with achromatic subwavelength architectures for light filtering and advanced SERS detection. ACS Applied Materials & Interfaces. 2016;**8**:8024-8031

[73] Karadan P, Anappara AA, Moorthy V, Narayana C, Barshilia HC. Improved broadband and omnidirectional light absorption in silicon nanopillars achieved through gradient mesoporosity induced leaky waveguide modulation. RSC Advances. 2016;**6**:109157-109167

[74] Turk N, Raza A, Wuytens P, Demol H, Van Daele M, Detavernier C, et al. Comparison of free-space and waveguide-based SERS platforms. Nanomaterials. 2019;**9**:1401

[75] Ross BM, Wu LY, Lee LP. Omnidirectional 3D nanoplasmonic optical antenna array via soft-matter transformation. Nano Letters. 2011;**11**:2590-2595

[76] Wuytens PC, Skirtach AG, Baets R. On-chip surface-enhanced

Raman spectroscopy using nanosphere-lithography patterned antennas on silicon nitride waveguides. Optics Express. 2017;**25**:12926-12934

[77] Chen Y-S, Ho H-C, Lai Y-C, Nagao T, Hsueh C-H. Thermochromic vanadium dioxide film on textured silica substrate for smart window with enhanced visible transmittance and tunable infrared radiation. Infrared Physics & Technology. 2019;**102**:103019

[78] Cheon SE, Lee Hs, Choi J, Jeong AR, Lee TS, Jeong DS, et al. Fabrication of parabolic Si nanostructures by nanosphere lithography and its application for solar cells. Scientific Reports. 2017;**7**:7336

[79] Mailoa JP, Lee YS, Buonassisi T, Kozinsky I. Textured conducting glass by nanosphere lithography for increased light absorption in thin-film solar cells. Journal of Physics D: Applied Physics. 2014;**47**:085105

[80] Kayes MI. Micro/Nano Structured Materials for Enhanced Device Performance and Antibacterial Applications. Univ. Pittsburgh; 2019

[81] Wattanatorn N. Nanosphere Lithography for Intracellular Delivery. Los Angeles: University of California; ProQuest Dissertations Publishing. 2019. 13899730

Chapter 8

Fabrication of Nanowire Arrays CuO-Al_2O_3-TiO_2 as p-Insulator-n Heterojunction for Photochemical Water Splitting

Yu-Min Shen, Dipti Ranjan Sahu, Sheng-Chang Wang and Jow-Lay Huang

Abstract

CuO@TiO_2 nanowires were prepared with the use of a porous alumina membrane (PAM). The conventional thermal chemical vapor deposition (CVD) method was used for the deposition of TiO_2 by decomposing titanium isopropoxide ($Ti(OiPr)_4$). The multilayer heterojunction is tested for possible photochemical water splitting application. The photo conversion properties of the CuO-Al_2O_3-TiO_2 (p-insulator-n) heterojunction along with the microstructures and composition were characterized by Potentiostat, SEM and TEM, respectively. The shape of CuO nanowire arrays were hexagonal honeycombs and size was about 90 nm which depends on the uniform pore size of the PAM. The microstructural characterization showed that the nanosized CuO-Al_2O_3-TiO_2 is a p-insulator-n heterojunction. The maximum photoconversion efficiency of 1.13 and 1.61% is obtained for CuO and CuO-Al_2O_3-TiO_2 nanowire arrays junctions. An energy band diagram was introduced to explain the change in current during water splitting due to electron tunneling through the insulating layer.

Keywords: nanowires, porous alumina membrane, TiO_2, CuO and photoelectrochemical, chemical vapour deposition (CVD)

1. Introduction

The use of TiO_2 for photoelectrochemical (PEC) water splitting applications has been attracted a great deal attention from many research groups [1–8] since Fujishima et al. [9, 10] discovered that the hydrolysis of water in oxygen and hydrogen could be carried out by lighting anatase phase of titanium oxide (TiO_2). This is because TiO_2 has a wide band gap of ~3.2 eV, appropriate band-gap positions, strongly optical adsorption, and high chemical stability. However, the conversion efficiency of TiO_2 is usually limited, due to large band-gap energy, which means that UV irradiation is need for photocatalytic activity to occur. Since UV light accounts for only 5% of the sun's energy compared to the 45% of visible, enhancing the photoconversion efficiency of TiO_2 has been a key objective of many research teams. H_2 has been used to treat TiO_2 in order to modify its optical absorption responsive band gap from UV to the visible wavelength [8, 11, 12], and narrow band gap semiconductor have also been used to avoid the problem of rapid electron-hole recombination [13–17]. For example, it is known that p-type copper oxide (CuO) can act as a hole conductor due to its narrow bandgap of only 1.4 eV. The use of copper oxides heterojunctions with TiO_2 have been reported for enhanced hydrogen generation, due to the resulting shift in the adsorption light wavelength [18].

p-n heterojunctions with various structures have been developed for photoconversion applications, resulting in high photocurrent efficiencies and H_2 generation [19–27]. The interface layers between p-type and n-type heterojunction semiconductors have significant effects on the resulting opto-electric properties, which are enhanced when the electrons at the conduction band (CB) were rapidly move from the n-type semiconductor to p-type semiconductor, or vice versa, which depends on the semiconductor band edge position. The interfaces of p-n heterojunction have recently been shown to have excellent photovoltaic properties, because the intermediate intrinsic layer prevents the injection of holes from the p-type semiconductor [28, 29]. Tian et al. [30] also indicated that p-type/intrinsic/n-type coaxial silicon nanowires have a high threshold voltage, as seen in the result of I-V characterization, due to the contributions from tunneling and avalanche mechanisms [31]. However, the influence of an insulator interface between p-n heterojunctions on the resulting water splitting behavior has not been examined.

An insulator layer between p-type and n-type semiconductors was prepared for this study by electrochemical deposition (ECD) and chemical vapor deposition (CVD), for the fabrication of the heterojunction structure. In the ECD process, the CuO/Al_2O_3 nanowire arrays were prepared with the use of template assistance process [32, 33]. The TiO_2 particles were deposited by a thermal CVD process. An energy band structural diagram is used to explain the I-V characteristics of the resulting, $CuO/Al_2O_3/TiO_2$ (p-insulator-n) heterojunction for water splitting applications.

2. Experimental methods

2.1 Preparation of porous alumina membranes (PAMs)

The porous alumina membranes (PAM) are prepared using high purity Al foils (99.9995%) with two step anodization process [34, 35]. The Al foil was degreased in ethanol, annealed in an argon, cleaned and electro-polished in a mixture of $HClO_4$-C_2H_5OH (1:4 vol %) at 10°C, with a current density of 100 mA/cm^2 for 1 min. The anodization was carried out in 0.3 sulfuric acid and 0.3 M oxalic acid at a constant voltage of 60 V using Pt foil as a counter electrode. After 3–6 h of anodization, the alumina film was selectively etched in a mixture of H_3PO_4-CrO_3-H_2O (2 g-3.5 mL-100 mL) at 70°C for 40 min. Afterwards second anodization was done under the same conditions for 18–24 h. The straight nano channel pore was obtained by dissolving the barrier layer using 5% H3PO4 solution at 60°C for 20 min.

2.2 Preparation of CuO/PAMs nanowire arrays

Before the deposition of the nanowire arrays, layers of Ti (10 nm) and Pt (100 nm) were coated onto one side of the membrane and stuck on the Cu substrate to serve as the working electrode (cathode). Pt foil (anode) and a saturated calomel electrode (SCE, 0.241V/NHE) served as the counter and reference electrode, respectively, in a three-electrode electrochemical deposition system. A potentiostat (263A, Princeton Applied Research) was used for electrochemical system to provide the potential. The PAMs/Ti/Pt/Cu were degassed using a mixture of 0.2 M $CuSO_4$ and 0.15 M H_3BO_3 solution with pH = 2 (controlled by H_2SO_4) and deposited at −0.18 V/SCE for 2 h. After electrochemical deposition, the Cu/PAMs nanowires arrays were heated in air at 400°C for 12 h to obtain the CuO/PAMs nanowires arrays.

2.3 Preparation of CuO@TiO_2 heterojunction structure by CVD

Thermal chemical vapor deposition (CVD) was used to grow the TiO_2 coatings on the CuO nanowires. Before the coating process, the CuO/PAMs compound was stuck on the FTO glass using silver paste, and the PAM was removed by 1 M NaOH at 50°C for 5 min. The CuO/PAMs sample was placed on the graphite holder in the quartz tube under reduced pressure of 10^{-5} Torr. The growth parameter was set at 500°C for 15 min. The precursor, titanium isopropoxide ($Ti(O\textit{i}Pr)_4$),was introduced to decompose the TiO_2 in the quartz tube at 35°C.

2.4 Characterization of the CuO/PAMs nanowires arrays and CuO@TiO_2

The CuO/PAMs was immersed into the 1 M NaOH solution at 50°C for 5 min, and then cleaned with deionized water several times before the SEM observation. The TEM sample for CuO@TiO_2 was dispersed in ethanol and subjected to ultrasonication for 10 min, then dropped on the Ni grid. To examine the photochemical response of CuO/PAM and CuO@TiO_2 p-insulator-n junction, sliver wire was stuck on the substrates, and this was then covered with a non-transparent and non-conductive epoxy-resin. The photochemical current-voltage was characterized using a potentiostat, while the counter and reference electrode were platinum and saturated calomel (SCE, 0.242V) in 1 M NaOH solution. Sunlight was simulated using a 150 W xenon arc lamp (1.5 AM filter, Oriel).

3. Results and discussion

3.1 Crystal structure and microstructure observation

Figure 1 presents the X-ray diffraction (XRD) spectra of the Cu/CuO nanowire arrays obtained after 400°C heat treatment for 12 h. It clearly shows the CuO peaks which correspond to the JCPDF card number 45-0937. The remaining Cu was due to the alumina template covering the surface of the Cu. The alumina template made it difficult for the oxygen to diffuse into it and to form CuO. **Figure 2(A)** presents the ordered pore array of AAO with 90–100 nm pore size obtained using two step anodination of Al foil under oxalic acid at 60 V [35]. The uniform shape of CuO nanowires were synthesized with the use of porous alumina membranes. **Figure 2(B)** shows the CuO nanowires after removing the PAM with NaOH solution. It is observed that the CuO nanowires were agglomerated and collapsed on both sides of AAO due to surface interaction. The SEM in **Figure 2(C)** represents the images of TiO_2 particles attached to the CuO nanowire after the CVD process. **Figure 2(D)** indicates the clear view of formation of nanosize CuO nanowires on PAM.

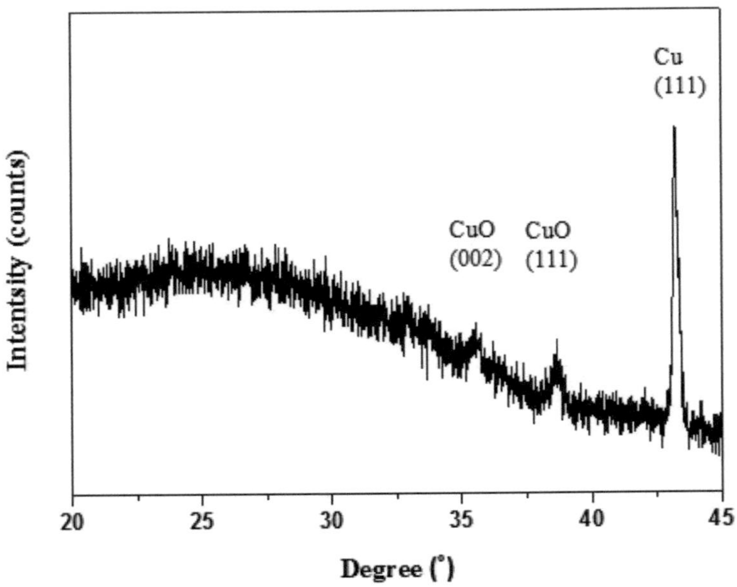

Figure 1.
XRD pattern of Cu/CuO after 400°C heat treatment for 12 h.

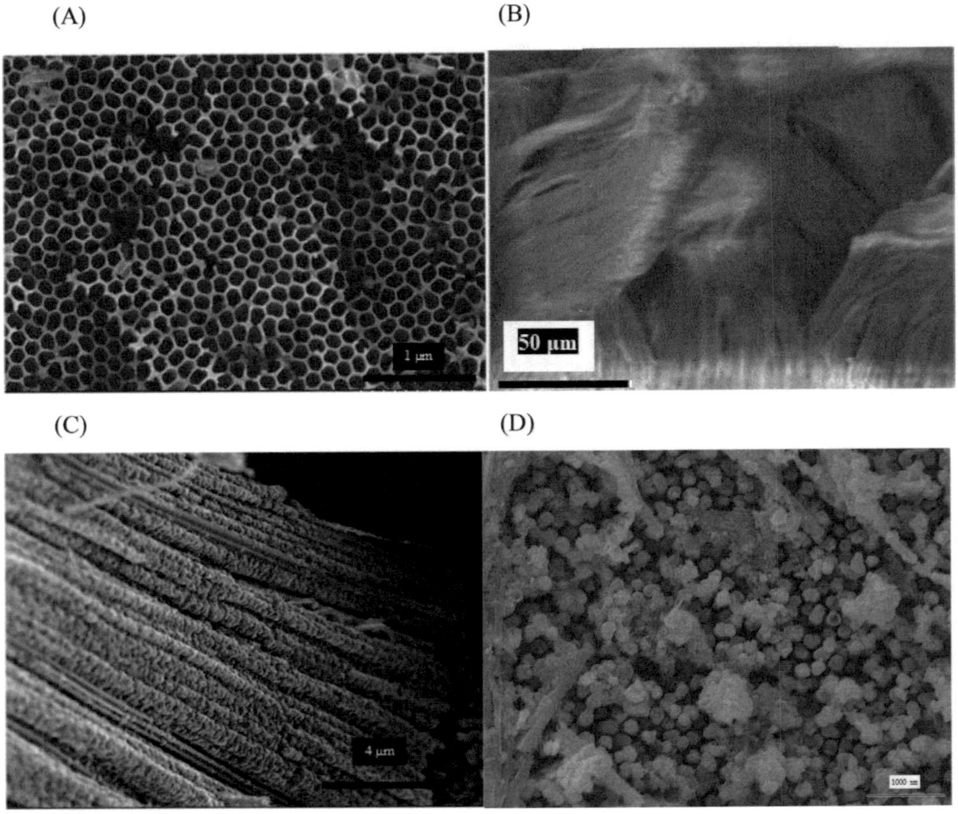

Figure 2.
SEM images of (A) porous alumina membranes, (B) CuO nanowire arrays, and (C) CuO@TiO_2 nanostructure, (D) clear view of nanosize CuO nanowire arrays.

Figure 3 shows the TEM, high-resolution TEM, and fast Fourier transformation (FFT) images of a CuO@TiO_2 nanowire. **Figure 3A** and **C** show that the particles cohered on the surface of nanowire. The d-spacings of the nanowire and particles were calculated as 0.242 nm and 0.446 nm, which confirm that the compounds were CuO (111) and TiO_2 (110), respectively. The high resolution TEM image shows that the CuO nanowire was a single crystal, and that its growth direction was (111).

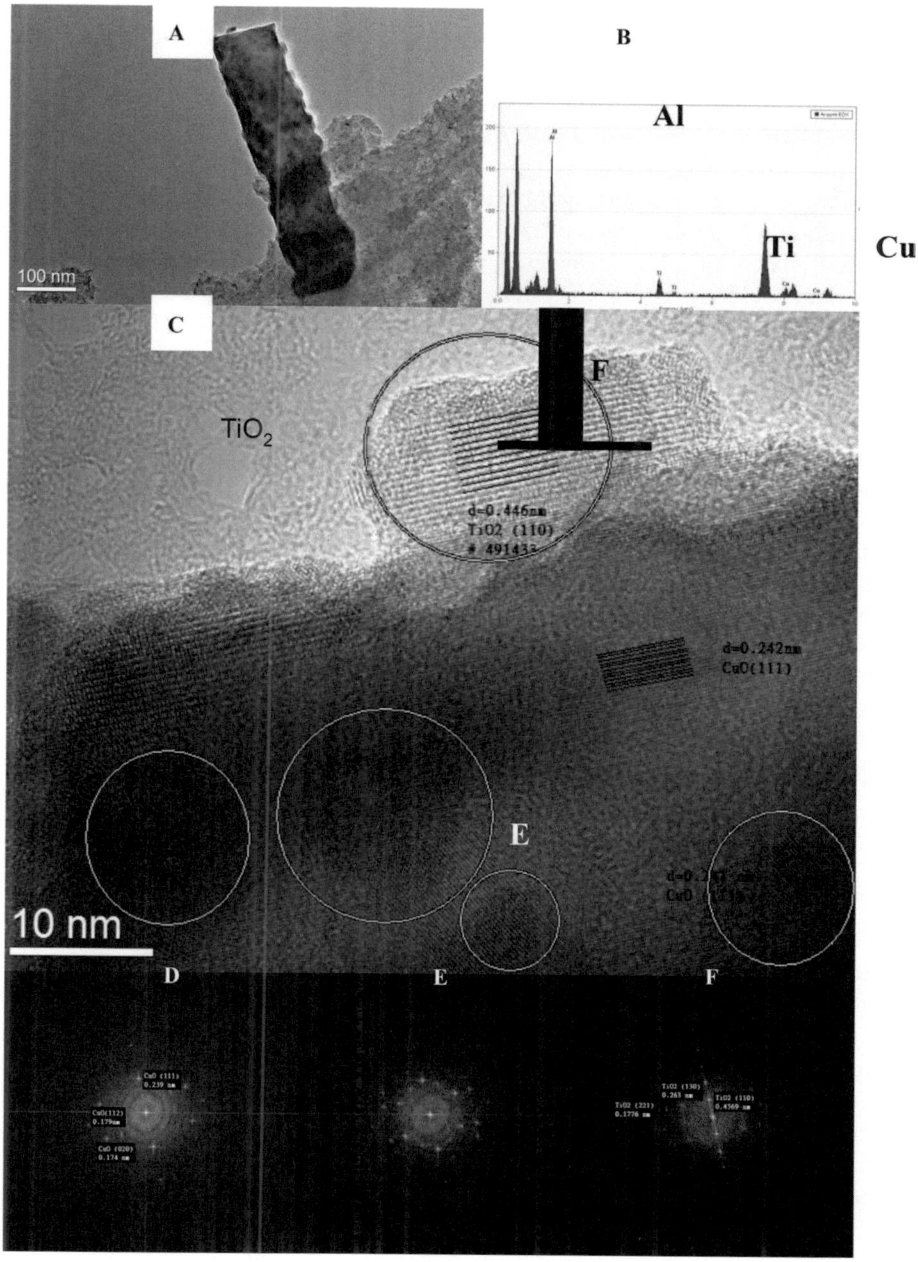

Figure 3.
TEM images of a CuO@TiO_2 nanowire (A) at low magnification, (B) EDS analysis of a CuO@TiO_2 nanowire, (C) high resolution, and fast Fourier transformation images, (D) CuO/PAMs, (E) CuO@TiO_2, and (F) TiO_2 particles produced by thermal CVD.

In addition, the CuO nanowire, $CuO@TiO_2$ and TiO_2 particles are also shown in the fast Fourier transformation images in **Figure 3C–E**. The impurity spots in **Figure 3D** show that the CuO nanowire was covered with TiO_2 particles. However, some indistinct regions can also be seen on the CuO nanowire in the HR-TEM and hollow ring in FFT images (**Figure 3C** and **D**), and this was due to the amorphous porous alumina membranes. **Figure 3B** shows the EDS analysis of a $CuO@TiO_2$ nanowire, which indicates the presence of the Al_2O_3 compound. This demonstrates that the insulator layer, i.e., PAM, was not completely removed during etching by 1M NaOH solution. The TEM and EDS results indicated that a p (CuO)-insulator (Al_2O_3)-n (TiO_2) junction was successfully fabricated.

3.2 Water splitting properties of $CuO/Al_2O_3/TiO_2$ p-insulator-n heterojunction

According to the water splitting theory, the optimum energy bandgap of the semiconductor should be in the range of 1.7–2.0 eV and, the band edges should be straddle water potential of 1.23 V/NHE. However, the photon adsorption of CuO is 1.4 eV, which do not provide enough energy to split water into hydrogen and oxygen. The wide band gap of TiO_2 not only provides enough band edges to straddle the H_2O reduction potentials but forms the donor level in the forbidden band to enable electron transition [13]. The photoelectrochemical and photocurrent characterization of the CuO/PAMs nanowire arrays and $CuO@TiO_2$ nanostructure by thermal CVD were carried out based on their current density and voltage properties under dark and UV illumination. **Figure 4(A)** presents the current density measurement of nanowire arrays with various applied potential under 150 W UV illumination. The CuO/PAMs nanowire arrays indicates p-type semiconductor behavior due to the signature of negative current in the figure. There is also a current density peak between 0.2 and 0.7 V/SCE due to the oxidation of the remaining Cu to Cu^{2+} as the potential increases to +1 V/SCE.

Figure 4(B) reveals the photochemical response of $CuO@TiO_2$ nanowire arrays under 100 mW/cm^2 simulated solar illumination. The threshold voltages (V_{th}) of the CuO/PAMs and $CuO@TiO_2$ samples were 0.67 and 0.89 V/SCE, while the maximum photocurrent densities were 1.12 and 1.83 mA/cm^2 respectively.

In order to examine the photoconversion efficiency (η) of each sample, the efficiency equation is expressed as follow [36]:

$$\eta(\%) = \frac{J_{ph}\left[\left(1.23 - \left|V_{app}\right|\right)\right]}{I_0} \times 100 \qquad (1)$$

Where I_0 is the intensity of the incident light (100 mW/cm^2, V_{app} is the applied potential, V_{mea} is the measured potential (obtained with a potentiostat, and V_{oc} is the open circuit potential. **Figure 5** presents the photo conversion efficiency of CuO/PAMs and $CuO@TiO_2$ samples. The conversion efficiency of 1.61% at 0.8 V/SCE, was observed for $CuOs@TiO_2$ heterojunction which is higher than that of 1.13 % at 0.67 V/SCE for CuO/PAMs.

3.3 Interface behavior of the $CuO/Al_2O_3/TiO_2$ (p-insulator-n) heterojunction

Based on the photochemical properties outlined above, the results for the $CuO@TiO_2$ sample show that it has better performance than the CuO/PAMs sample [37]. Bandara et al. [13] reported that in a CuO-TiO_2 water system the electrons and holes are generated on the conduction band (CB) and valence band (VB) when the TiO_2 is

(A)

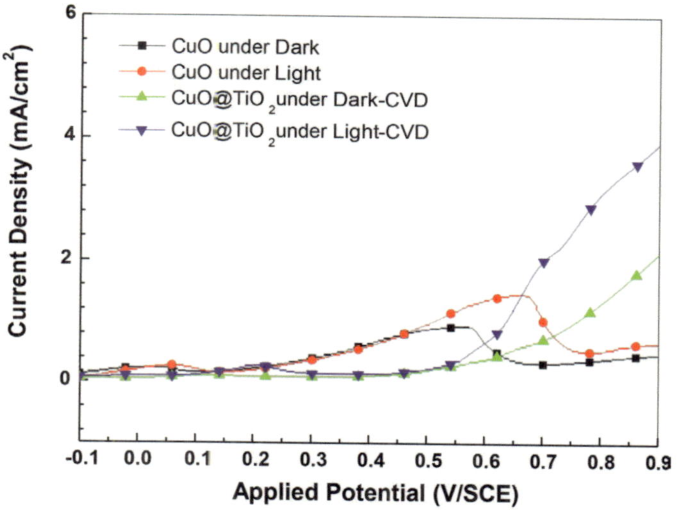

(B)

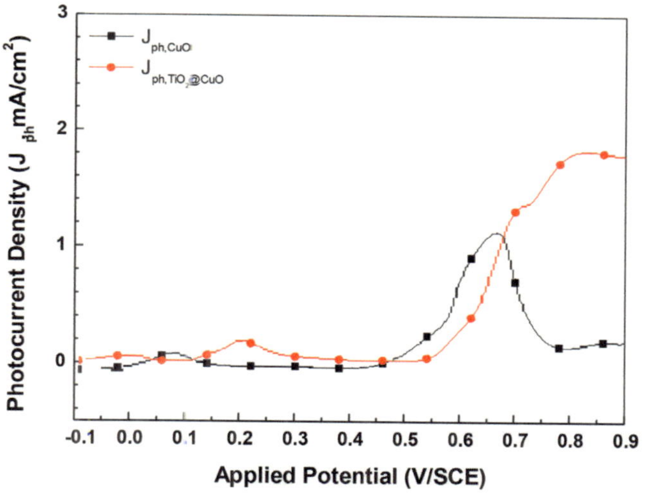

Figure 4.
(A) Variation of current density versus applied potential measured using 150 W UV illuminations (AM 1.5 filter) and (B) photocurrent density vs applied potential.

illuminated. The rapid electrons transfer from the CB of TiO_2 to the CB of CuO occurs because the CB position of CuO is lower than that of TiO_2. The threshold voltage (V_{th}) of the ideal CuO-TiO_2 p-n junction should be set at 0 V/SCE. However, in our study, the photo response of CuO@TiO_2 appeared at a relatively high V_{th} (0.9 V/SCE). This was due to the Al_2O_3 insulation layer between the CuO and TiO_2 interlayers, which formed the heterojunctions, as shown in **Figure 6**. The band gaps of CuO, Al_2O_3, and TiO_2 are 1.4, 8.8, and 3.2 eV, respectively. In this heterojunction, insulator (Al_2O_3) was used as an electron blocking layer. In order to balance Fermi level, (the interface between the p- and n-type) insulators started to bend when applying a bias due to the high band gap of this layer. This caused electrons to accumulate on the barrier

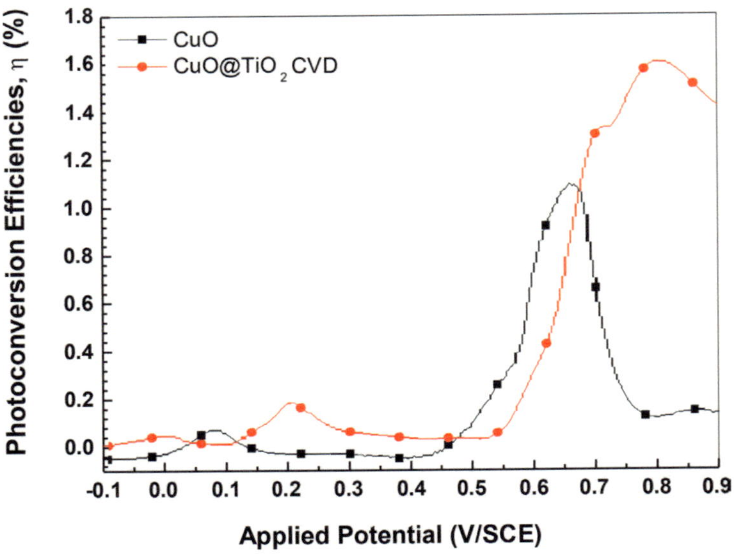

Figure 5.
Plot of photoconversion efficiency of versus applied potential of Cu/PAM and CuO@TiO_2 nanowire arrays.

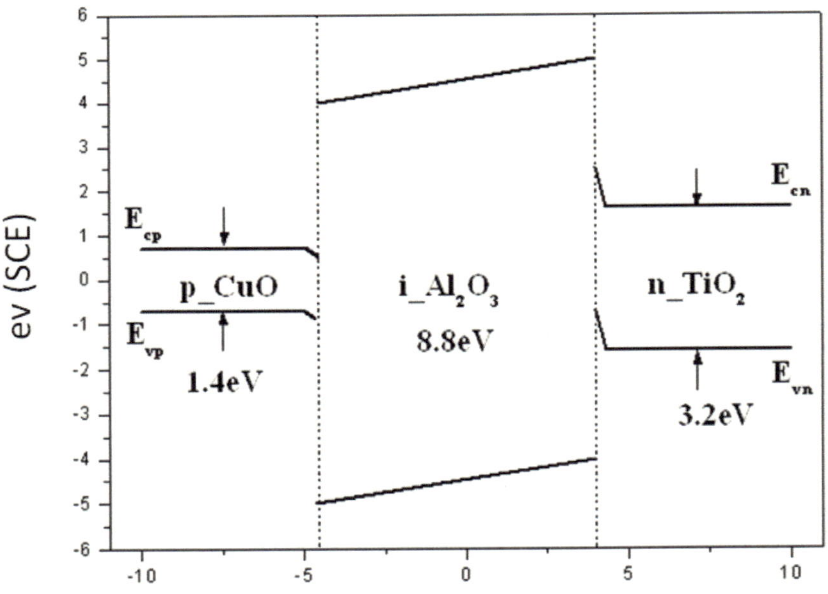

Figure 6.
Energy band gap structural diagram of CuO-Al_2O_3-TiO_2 heterojunctions.

layer before the breakdown potential was reached. When the applied potential was increased to the breakdown level, at ~0.6 V/SCE, the rapidly increasing current was caused by electron tunneling.In this CuO-Al_2O_3-TiO_2 heterojunction, the conduction band edge of CuO is more negative than TiO_2. Upon illumination the photogenerated electrons (e^-) and holes (h^+) are produced at the active sites of the CuO-TiO_2 photocatalyst. The electrons in p-type CuO easily flow to the conduction band of TiO_2, whereas the photogenerated holes migrate in the opposite direction. CuO serves as electron reservoir by receiving electrons from TiO_2, which suppresses the recombination of e^-/h^+ and transfers the received electron.

4. Application

CuO based photoelectrodes have great potential for applications in photo electrochemical (PEC) water splitting which give useful information on factors that affect the photo electrochemical capability. Modifying the CuO structure revealed a significant influence on the enhancement of the photocurrent density and photostability. Further, copper containing TiO_2 is cost-effective compared with noble metal loaded TiO_2, for the photocatalytic hydrogen production. The CuO-TiO_2 oxide system have promising photoactive optical and photocatalytic properties, as compared with TiO_2 or CuO alone. CuO and TiO_2 based junction have better charge mobility at the proximity to the junction of the electrode/electrolyte within the nanostructure. CuO-Al_2O_3-TiO_2 nanocomposites represent the synergy effect for hydrogen production. This nanocomposites junction is very good for design of efficient catalyst for selective alcohol conversion to other valuable chemicals. The p-i-n heterojunctions with suitable band edge positions has the improved separation of the photogenerated charge which improved the photocatalytic activity. This junction has application for improved photocurrent behavior, the H_2 generation rate by water splitting in a full PEC device (without application of a bias) and the solar-to-hydrogen efficiency.

5. Conclusions

CuO@TiO_2 nanostructures were fabricated by a template assisted (PAMs) nanowires (CuO) growth and CVD process (TiO_2). Observations of the microstructure show that the CuO@TiO_2 nanostructure was composed of p-insulator-n heterojunctions. The I-V characteristics of the CuO/PAMs and CuO@TiO_2 nanostructures show high threshold voltages of 0.67 and 0.89 V/SCE, with photoconversion efficiencies of 1.13% and 1.61%, respectively. The energy band structure diagram indicates that the rise in current during water splitting is due to electron tunneling through the insulation layer (Al_2O_3).

Acknowledgements

This work was supported by the National Science Council of Taiwan under grant NSC 101-2221-E-006-129.

Author contributions statement

DRS, SCW, JLH conceived the experiment(s), YMS conducted the experiment(s) and characterization. YMS and DRS wrote the manuscript. YMS, DRS and SCW analyzed the results.

All authors reviewed the manuscript.

Competing financial interests

The authors declare no competing financial interests.

Author details

Yu-Min Shen[1], Dipti Ranjan Sahu[2*], Sheng-Chang Wang[3] and Jow-Lay Huang[1,4,5*]

1 Department of Materials Science and Engineering, National Cheng Kung University, Tainan, Taiwan

2 Faculty of Health, Natural Resources and Applied Sciences, School of Natural and Applied Sciences, Department of Biology, Chemistry and Physics, Namibia University of Science and Technology, Windhoek, Namibia

3 Department of Mechanical Engineering, Southern Taiwan University of Science and Technology, Tainan, Taiwan

4 Research Center for Energy Technology and Strategy, National Cheng Kung University, Tainan, Taiwan

5 Hierarchical Green-Energy Materials (Hi-GEM) Research Center, National Cheng Kung University, Tainan, Taiwan

*Address all correspondence to: dsahu@nust.na; jlh888@mail.ncku.edu.tw

References

[1] Mor GK, Shankar K, Paulose M, Varghese OK, Grimes CA. Enhanced photocleavage of water using titania nanotube arrays. Nano Letters. 2005;**5**:191-195

[2] Wang D, Zhang X, Sun P, Lu S, Wang L, Wang C, et al. Photoelectrochemical water splitting with rutile TiO_2 nanowires array: Synergistic effect of hydrogen treatment and surface modification with anatase nanoparticle. Electrochimica Acta. 2014;**130**:290-295

[3] Kochuveedu ST. Photocatalytic and photoelectrochemical water splitting on TiO_2 via photosensitization. Journal of Nanomaterials. 2016;**4073142**:1-12

[4] Li F, Huang Y, Peng H, Cao Y, Niu Y. Preparation and photocatalytic water splitting hydrogen production of titanium dioxide nanosheets. International Journal of Photoenergy. 2020;**3617312**:1-6

[5] Wang G, Wang H, Ling Y, Tang Y, Yang X, Fitzmorris RC, et al. Hydrogen-treated TiO_2 nanowire arrays for photoelectrochemical water splitting. Nano Letters. 2011;**11**:3026-3033

[6] Yew R, Karuturi SK, Liu J, Tan HH, Jagadish C. Exploiting defects in TiO_2 inverse opal for enhanced photoelectrochemical water splitting. Optics Express. 2019;**27**:761-773

[7] Joy J, Mathew J, George SC. Nanomaterials for photoelectrochemical water splitting-review. International Journal of Hydrogen Energy. 2018;**43**:4804-4817

[8] Peng C, Wang W, Zhang W, Liang Y, Zhou L. Surface plasmon-driven photoelectrochemical water splitting of TiO_2 nanowires decorated with Ag nanoparticles under visible light illumination. Applied Surface Science. 2017;**101**:27-33

[9] Fujishima A, Honda K. Electrochemical photolysis of water at a semiconductor electrode. Nature. 1992;**238**:37-38

[10] Fujishima A, Kohayakawa K, Honda K. Hydrogen production under sunlight with an electrochemical photocell. Journal of the Electrochemical Society. 1975;**122**:1487-1489

[11] Zhang XY, Li HP, Cui XL, Lin Y. Graphene/TiO_2 nanocomposites: Synthesis, characterization and application in hydrogen evolution from water photocatalytic splitting. Journal of Materials Chemistry. 2010;**20**:2801-2806

[12] Pu YC, Wang G, Chang KD, Ling Y, Lin YK, Fitzmorris BC, et al. Au nanostructure-decorated TiO_2 nanowires exhibiting photoactivity across entire UV-visible region for photoelectrochemical water splitting. Nano Letters. 2013;**13**:3817-3823

[13] Bandara J, Udawatta CPK, Rajapakse CSK. Highly stable CuO incorporated TiO_2 catalyst for photocatalytic hydrogen production from H2O. Photochemical & Photobiological Sciences. 2005;**4**:857-861

[14] Zhang YG, Ma LL, Li JL, Yu Y. In situ Fenton reagent generated from TiO_2/Cu2O composite film: A new way to utilize TiO_2 under visible light irradiation. Environmental Science & Technology. 2007;**41**:6264-6269

[15] Mor GK, Varghese OK, Wilke RHT, Sharma S, Shankar K, Latempa TJ, et al.

p-type Cu–Ti–O nanotube arrays and their use in self-biased heterojunction photoelectrochemical diodes for hydrogen generation. Nano Letters. 2008;**8**:1906-1911

[16] Helaïli N, Bessekhouad Y, Bouguelia A, Trari M. Visible light degradation of. Orange II using xCu*y*Oz/TiO_2 heterojunctions. Journal of Hazardous Materials. 2009;**168**:484-492

[17] Tsai TY, Chang SJ, Hsueh TJ, Hsueh HT, Weng WY, Hsu CL, et al. p-Cu2O-shell/n-TiO_2-nanowire-core heterostucture photodiodes. Nanoscale Research Letters. 2011;**6**:575

[18] Liu Z, Bai H, Xu S, Sun DD. Hierarchical CuO/ZnO "corn-like" architecture for photocatalytic hydrogen generation. International Journal of Hydrogen Energy. 2011;**36**:13473-13480

[19] Mor GK, Prakasam HE, Varghese OK, Shankar K, Grimes CA. Vertically oriented Ti–Fe–O nanotube Array films: Toward a useful material architecture for solar spectrum water photoelectrolysis. Nano Letters. 2007;**7**:2356-2364

[20] Musselman KP, Marin A, Schmidt-Mende L, MacManus-Driscoll L. Incompatible length scales in nanostructured Cu2O solar cells. Advanced Functional Materials. 2012;**22**:2202-2208

[21] Deo M, Shinde D, Yengantiwar A, Jog J, Hannoyer B, Sauvage X, et al. Cu2O/ZnO hetero-nanobrush: Hierarchical assembly, field emission and photocatalytic properties. Journal of Materials Chemistry. 2012;**22**:17055-17062

[22] Jiang T, Xie T, Zhang Y, Chen L, Peng L, Li H, et al. Photoinduced charge transfer in ZnO/Cu2O heterostructure films studied by surface photovoltage technique. Physical Chemistry Chemical Physics. 2010;**12**:15476-15481

[23] Wei H, Gong H, Wang Y, Hu X, Chen L, Xu H, et al. Three kinds of Cu2O/ZnO heterostructure solar cells fabricated with electrochemical deposition and their structure-related photovoltaic properties. CrystEngComm. 2011;**13**:6065-6070

[24] Wang G, Yang X, Qian F, Zhang JZ, Li Y. Double-sided CdS and CdSe quantum dot Co-sensitized ZnO nanowire arrays for photoelectrochemical hydrogen generation. Nano Letters. 2010;**10**:1088-1092

[25] Kim H, Seol M, Lee J, Yong K. Highly efficient photoelectrochemical hydrogen generation using hierarchical ZnO/WO*x* nanowires cosensitized with CdSe/CdS. Journal of Physical Chemistry C. 2011;**115**:25429-25436

[26] Gutsche C, Lysov A, Braam D, Regolin I, Keller G, Li ZA, et al. n-GaAs/InGaP/p-GaAs core-multishell nanowire diodes for efficient light-to-current conversion. Advanced Functional Materials. 2012;**22**:929-936

[27] Zhu T, Chong MN. Prospects of metal–insulator–semiconductor (MIS) nanojunction structures for enhanced hydrogen evolution in photoelectrochemical cells: A review. Nano Energy. 2015;**12**:347-373

[28] Zhang DK, Liu YC, Liu YL, Yang H. The electrical properties and the interfaces of Cu2O/ZnO/ITO p-i-n junction. Physics B. 2004;**351**:178-183

[29] Ly P, Zheng W, Lin L, Peng F, Huang Z, Lai F. *I–V* characteristics of ZnO/Cu2O thin film n–i–p

heterojunction. Physics B. 2011;**406**:1253-1257

[30] Tian B, Zheng Z, Kempa TJ, Fang Y, Yu N, Yu G, et al. Coaxial silicon nanowires as solar cells and nanoelectronic power sources. Nature. 2007;**449**:885-889

[31] Hayden O, Agarwal R, Lieber CM. Nanoscale avalanche photodiodes for highly sensitive and spatially resolved photon detection. Nature Mater. 2006;**5**:352-356

[32] Shen YM, Shih YT, Wang SC, Nayak PK, Huang JL. Characterization of ordered Cu_2O nanowire arrays prepared by heat treated Cu/PAM composite. Thin Solid Films. 2010;**519**:1687-1692

[33] Shen YM, Pan CH, Wang SC, Huang JL. Ordered ZnO/AZO/PAM nanowire arrays prepared by seed layer assisted electrochemical deposition. Thin Solid Films. 2011;**520**:1532-1540

[34] Tan M, Chen X. Growth mechanism of single crystal nanowires of fcc metals (Ag, Cu, Ni) and hcp metal (Co) electrodeposited. Journal of Electrochemical Society. 2011;**159**:K15-K20

[35] Shen YM, Chiu WF, Wang SC, Nayak PK, Sahu DR, Huang JL. Variation of crystallinity of Cu and Cu2O nanowires arrays grown in various pores of porous alumina membrane. Advanced Materials Letters. 2017;**8**(11):1046-1051

[36] Aljohani TA, Almutairi AK. High photo conversion efficiency obtained from novel TiO_2 photoanodes. International Journal of Electrochemical Science. 2016;**11**:6848-6861

[37] Shen YM, Fiz R, Mettenborger A, Singh AP, Wang SC, Mathur S, et al. Characterization of $CuO-Al_2O_3-TiO_2$ p-i-n heterojunction nanowire arrays for photochemical water splitting. In: IEEE 8th Nanotechnology Materials and Devices Conference Presentation; IEE Nanotechnology Council, Tainan, Taiwan. 6-9 October, 2013